New House/More House

New House/More House

Solving the Residential Construction Project Puzzle

Richard Preves, AIA

Portico Publishing, LLC

ISBN: 0-9711044-0-9

Library of Congress Control Number: 2001-132443

Cover design: Pearl & Associates

Typesetting, design and production: Tabby House

Printed in the United States of America

Photos courtesy of Frank Merrill Photography

The AIA contracts in the appendix are reprinted by permission of the American Institute of Architects

Publishers Cataloging in Publication
(Provided by Quality Books, Inc.)

Preves, Richard
    New house/more house : solving the residential
construction project puzzle / Richard Preves
    p. cm.
    includes index
    LCCN: 2001132443
    ISBN: 0-9711044-0-9

    1. House construction--Popular works.   1. Title

TH4811.P74 2001   690'.837
                           QB101-200803

Portico Publishing, LLC
704 Florsheim Drive
Libertyville, Illinois 60048
www.newhousemorehouse.com.

This book is dedicated to my mother and in memory of my father, who encouraged me to explore any subject that caught my interest.

Acknowledgments

My appreciation extends to many people who helped make this book possible. To Dan Friedlander, who guided the preparation of the original set of newspaper articles and lectures. To Martha Brown, who helped organize and edit the materials created by an architect who is not a natural writer. Special thanks to Tim Bennett for his graphics help and ideas. To Don and Norma Crossett for their kind permission for the use of the house plans and model examples. To Vicky Buttitta and Khahn Phan for their efforts and a special thanks to Jean Quill for her enthusiasm and limitless efforts in producing the manuscript.

Finally, my warmest gratitude to my wife, Sandy, whose constant encouragement, help and understanding enabled me to complete this project.

Contents

Part One
What You Need to Know

Part Two
What You Need to Do

Part Three
Remodeling and Adding on to Your House—What You Need to Know and Do

Introduction

Making a living in the building industry is a study of contrasts. In booming economic times you can find plenty of work. During recessions, projects can be few and far between. Architects and builders know the dynamics of time; either you often feel helpless trying to meet a never-ending string of deadlines, or there is plenty of time to perfect your golf game.

If the second half of the 1980s were very kind to our fellowship, the early 1990s were brutal. As an architect and former general contractor, I was spending my time trying to re-invent myself to find more work. One client, who was a developer hard-hit by the recession had a favorite saying, "Oh Lord, grant me one more real estate boom and I promise I won't waste it away!"

Since the commercial side of my practice was slow, I decided to concentrate my efforts on increasing my custom residential practice, a sector that seemed to still be active in the Chicago area. Obtaining good house commissions is usually a matter of name recognition and word-of-mouth. You have to get your name on the street to succeed. Skipping the cocktail party circuit, I developed a program of writing articles for local newspapers on residential architecture and construction. In exchange for providing a free writing service, the newspapers gave me 500 words and my picture every two weeks.

From my past experience, topics for these articles were abundant. I have always concluded that most homeowners entering a residential construction project have little idea what they will encounter. A good comparison is either going to see a doctor or taking your car to a mechanic. Because you have no way of evaluating whether they are right or wrong, you must blindly trust their judgement.

Since my practice has always included both residential and commercial projects, it has been my policy to apply the same level of professionalism to both large corporations as well as homeowners. Many of the horror stories that abound regarding residential projects boil down to a lack of attention on both the part of the building professionals and the homeowner. So I decided to concentrate my articles on giving advice to prospective homeowners to avoid becoming the subject of yet another disastrous project. I expanded my writing to giving free lectures at public libraries, which were usually well received. It was at this point I discovered that homeowners were more interested in what I was saying than hiring me to do potential projects. The information I developed was more valuable than the fees I was trying to generate through the articles and lectures.

Thus *New House/More House* was created as a culmination of the articles, lectures and later, a seminar. I have structured the book in the same manner I guide my clients through a project. Over the years, I have found that homeowners' lack of knowledge falls in fairly predictable categories and problems that usually arise can be anticipated and avoided.

Upon reflection of the usual problems one confronts or hears of, I have also concluded that the leading factor in determining the success or failure of a project is the prospective homeowner. I contend that a direct correlation exists between the organizational skills of a homeowner and the success of a home construction project. Organized people usually produce a successful project; disorganized people usually produce disasters. Don't despair! If you are disorganized by nature, I will help you focus your time and energies to make you more efficient.

For most of us, our house is the largest financial investment of a lifetime. Whether you are building a new residence, adding on or remodeling your present home, a large amount of money is at stake. I have had friends who spent more time researching the purchase of a $40,000 car than they did spending $400,000 building a new house. Before buying the car, they checked the track record for performance and repairs of several models, made several test drives, and spoke with other owners for their opinions. In other words, they did their homework. When it came to constructing their new house and spending ten times that amount, they trusted the first builder they met. Trust in itself is a necessary commodity, but before you can trust, you must have confidence in those working for you. Confidence will be forthcoming based upon your knowledge of the process acquired from this book.

The residential construction industry can be a strange place for the layman to enter. I can think of no other field so large and yet contain so much difference in quality. I had a professor in architectural school that once tried to explain this phenomenon. He told us that building a custom home project could also be compared to purchasing a car. The prospective owner visits the offices of General Motors, Ford and Chrysler for the purpose of selecting a car builder. After choosing one company, he has frequent meetings to determine the design of the car. At the conclusion of the planning stage, the car company arrives one day at his home and proceeds to build the car on his driveway, assembling all the parts as they arrive. Battling weather, delays in material delivery and labor shortages, the car is finally completed late and over budget, in part due to the owner changing the upholstery color twice and deciding after it was installed, that the stick-shift transmission had to be changed to automatic! Now would anyone in his right mind buy a car this way? Yet this is precisely the way homes have been constructed since man decided to live within walls and under a roof!

You will find MBAs running construction firms handling many projects all the way down to small contractors operating out of a pickup truck. To whom would you choose to entrust your project? The obvious is not always correct. It has been my experience that on the right project, the small contractor can run rings around his more professionally trained counterpart. With these types of inconsistencies, the homeowner must be schooled in the art of evaluation.

In *New House/More House*, I present a systematic six-step program providing the perspective homeowner with valuable information, enabling them to make intelligent, informed decisions at those critical forks in the road. There should be no reason why a residential construction project cannot be enjoyable, after-all, the choice of shaping one's living environment is a luxury not shared by all.

Our program, with diagrams, examples, and forms is applicable for new houses, whether custom or tract, additions and remodelings. Depending upon the size of your project, you may feel we are trying to kill an ant with an atomic bomb. The favorite phase you often hear from certain building personalities is, "That's only used on big jobs, we have never needed to do that before." As we are about to begin, a word to the wise: only you can control your project's path. Small projects can generate as many headaches as the big ones.

There is nothing I have yet found to compare with the feeling of pride and accomplishment on the part of a homeowner involved in a successful project. The emotions of dreams, happiness and security play a major part in creating your living environment. We want to make it a positive experience for you to enjoy and look back upon with pleasant memories.

How to Use This Book

New House/More House is divided into three separate sections and is sequential in nature, much like a construction project. Part 1, titled "What You Need to Know," forms the foundation for subsequent parts to be built upon. Chapters 1 through 6 introduce concepts to familiarize you with the project process, the members of the residential building industry, and terms and procedures. While providing a good deal of specific information, Part 1 also produces a lot of questions.

These questions are answered in detail in Part 2, "What You Need to Do." Chapters 7 through 12 escort you through the project process, pointing out the possible scenarios you could encounter and our strategies to optimize your outcome.

While Parts 1 and 2 are meant to be applied to new houses, additions and remodelings, Part 3, "Remodeling and Adding On to Your House, What you Need to Know and Do" addresses issues specific to additions and remodelings. This information is an extension of chapters 1 through 12, and builds upon the concepts previously introduced.

One of the book's major themes is that a successful project is directly related to the amount of knowledge you possess. You will receive the maximum benefit reading the entire book in sequence, as it closely parallels the course of a typical project.

For those of you who do not read "help books" cover to cover, and I include myself in this group, chapter 1 provides a good, short summary of the core concepts, and will supply you with the basics. I have also included at the end of each chapter, a recap of the important points introduced and discussed.

Before we begin, one word of caution. The purpose of this book is to provide the reader with general information regarding the topics covered. Because every project is unique, *New House/More House* should not be regarded as providing opinion or advice for any individual project. The author is not engaging in rendering specific architectural, legal, or other professional services. If architectural, legal or other professional advise or expertise is required, the reader should seek the services of competent local professionals. Should you desire specific additional information for your project, visit our **Web site: www. newhousemorehouse. com**.

Finally, for convenience, I often refer to the building professionals by using masculine pronouns. This is not intended as a slight, as the industry abounds with female participants.

Part One

What You Need to Know

Chapter 1

The Six-Step Program for Project Success

Years ago in the course of my architectural practice, I was called to visit a very expensive new home built for a doctor and his family. He showed me the basement, where I saw huge cracks in the concrete foundation walls and floors, permitting underground water to enter. The doctor was at his wit's end. The general contractor had made some halfhearted attempts to correct the problem, but water still seeped into the house every time it rained.

The doctor proceeded to tell me his hard-luck story. He had selected a very reputable local architect to design his dream house and then obtained four competitive bids from qualified general contractors. Assuming he would receive extra value, service, and workmanship from the highest bidder, he chose the most expensive contractor's bid, nearly $75,000 above the next highest bid. "I thought the highest bidder would ensure the best construction," he said, "How could I be so wrong?"

The doctor made a classic mistake that is at the root of many a construction horror story. Instead of taking the time to better understand the process he was involved in, he took what he perceived as a shortcut by hiring the highest bidder, an expensive shortcut that required spending even more money to fix the problem.

New House/More House presents a six-step program designed to escort the novice through the initiation rites of joining the construction fraternity. Everyone has to pay dues to join the club, and by following these six steps, you will take your own giant step toward ensuring that your project fulfills your dreams and objectives. After all, six steps is only four more than learning to dance the Texas Two-Step!

For those of you who don't have the time to read the book cover to cover, adopting some of these broadly brushed points as strategies will serve you in good stead. Each step is a miniature description of broad concepts discussed in great detail in the following chapters. If I seem to throw a lot of ideas at you at this stage, don't worry! We will spend "quality time" together so you will thoroughly understand each topic.

Step One: Learn the Process

Take the first step by becoming acquainted with the orderly sequence of a residential project. The design and construction process is very dependent on completing one task before starting the next. Just as you would never fly on an airplane that was untested, you also shouldn't tell the contractor to start construction without knowing where to start digging! A persistent problem is that building professionals, such as architects, builders and general contractors, often try to bypass or re-arrange the process for their convenience or profit. It is very important that you thoroughly understand each step of the process so you can have the confidence that your project is proceeding properly at all times.

Figure 1-1 (page 5) *Residential Project Sequence* presents a flowchart that walks you through the major steps of the entire process. As you scan down the sequence you are probably encountering phrases you have never seen before. Don't worry. Each of these activities will be explained in depth throughout *New House/More House*. What you need to know now is that a building project involves three distinct phases: Pre-Design, Design, and Construction.

Pre-Design consists of activities almost exclusively accomplished by you. In my experience gained over twenty-five years in the trenches, this phase is where a project's success is won or lost. Your project's outcome is directly proportionate to the amount of effort you put forth, particularly at the beginning. The fact that you are reading *New House/More House* shows you are interested in putting in a little extra effort.

Pre-Design actually starts when you get the idea that you wish to change your living environment. The lightbulb goes on! You need more space, you want to live in the country, or you won the lottery and the sky is the limit! Whatever the reason, you are visualizing something new and different. This is the most exciting phase of a project. Your imagination runs wild with ideas that know no limitations. Let 'er rip! Limitations such as space, time and budget will be forced on you all too soon in the process.

At this initial phase, many homeowners do not know how to verbalize their requirements, or whether they can afford their ideas. *New House/More House* introduces these issues and discusses them in depth in later chapters. You will learn how to articulate

your goals and requirements so you can more fully define for yourself and others what it is you are trying to accomplish. Once you have identified your needs, you have to decide whether your project is feasible. Can you really afford that new house, or is that addition physically possible? If you remodel one part of the house, what happens to the rest? Finally, and probably the most important, who is going to help you turn your ideas into physical reality? What do architects, builders, and general contractors really do? There are many players on the design and construction teams; you can't tell the players without a scorecard!

Once you have selected the professionals who will work for you, it is time to enter the *Design phase*. This is the classic picture of you and your designers meeting at the site or poring over piles of blueprints, discussing and developing different plans. Your initial ideas must be carefully evaluated at this stage to prioritize and separate out those dreams which may be a little beyond your reality.

Actually, the Design phase is largely about communication. You must be able to determine your goals and requirements for the project and then effectively convey them to your design professionals. They, in-turn, must be skilled in listening carefully to your thoughts and transforming them into three-dimensional concepts shown on drawings. This is another very critical point in your project; if you and your architect cannot communicate, your goals and dreams may not be realized.

Architectural design is a process that starts with broad, general ideas about the project and then proceeds by adding numerous specific details along the way. *Preliminary Design* starts the ball rolling by setting the project parameters and translating them into introductory plans, which show basic elements, such as exterior and interior shapes and spaces. Here your ideas spring into a tangible reality; you can begin to see walls, roofs, and rooms that you will

Figure 1-1

Residential Project Sequence

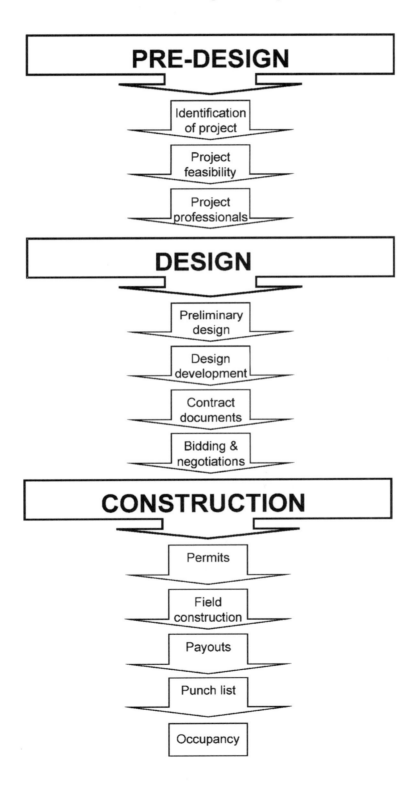

live within. But a lot of detail still remains to complete the design. *Design Development*, the second part, perfects the original design concept and selects the many elements that make up a house. Building materials, color, equipment and interior design details must be added to complete the design. Think of Design Development as a stage that was added to the process after man decided to move out of caves and live with more than just stone!

Once the design has been developed, all the information that has been compiled must be translated into a language that many participants in the construction industry can understand. *Construction Documents* are the complete set of drawings that the project is constructed from. They contain very specific information used by the dozens of material suppliers and tradesmen who need to understand your requirements and build the project. Although much of what you see on these drawings may appear to have been written in an alien language, every decision you make in the design process is contained within these sheets.

Before the first shovelful of dirt can be turned, a construction cost must be determined. During *Bidding and Negotiations*, the final step prior to construction, your budget and your project requirements are confirmed by the builder or general contractor. Here is another critical point where many projects hit their first major snag. *New House/More House* presents several strategies to help you avoid discovering that your completed design costs more than you have in your wallet.

The *Construction* phase is an exciting part of the process. Nearly everyday you can see your project take shape, with new materials and tradesmen arriving at the site to assemble your new home. Construction has many facets besides building the project. Prior to beginning construction, your team will need *Building Permits* for building, based on the contents of your drawings. As in almost any ac-

tivity today, our government has created a lot of red tape. In the construction industry, these regulations are known as building codes. These codes were developed over many years to protect public health and safety. Your drawings will be checked to see that your house project will have safe electrical work, sanitary plumbing and will stand up in the wind.

Field Construction is the process of building your house. You'll have two key concerns throughout construction: ensuring that your project is accurately built to meet your requirements and that the workmanship meets high standards. This is another point where horror stories originate. All the time you carefully spent planning the project can be compromised by the people you hire through sloppy workmanship and lack of attention to detail. *New House/More House* presents several methods to help you monitor construction progress and evaluate the project you see growing day by day.

With the tons of building materials being delivered to the site and skilled tradesmen measuring, cutting and installing, you will realize that a lot of money must be changing hands. Suppliers and workmen all want to be paid on a timely basis and are looking to you to have your checkbook readily available. That's why they treat you kindly at the site! *Payouts* are an intricate process during which you and your bank receive periodic requests for payment, evaluate all the paperwork, and cut the checks. *New House/More House* alerts you to the fact that your pocketbook is your most effective tool in managing the entire project. You should reward timely and accurate performance with prompt payment—and show your disapproval of slow progress and unacceptable work with incomplete payment.

The final two parts of the *Construction* phase, known as *Punch List* and *Occupancy*, help you close out your project successfully. Loose ends abound near the end of construction. You are eager to move in and enjoy the fruits of your labor. The building professionals also want to move on to their next assignment. But not so fast! Before everyone says good-bye, you must double-check that the construction work is complete and correct. Completing as much as possible prior to final payment means less disruption to you after you take occupancy. Many contractors or builders want you to occupy the project as soon as possible so they can receive their final monies and put off dealing with any problems in the house until the warranty period.

New House/More House discusses what a construction *Warranty* really does. No building project is ever 100 percent perfect. An effective warranty can help you identify these problem areas and get them fixed on a timely basis.

Step Two: Hire the Right Professionals

Hiring the right building professionals is the second step in our continuing process. Without a doubt, the most critical decision you must make is who will deliver your project. Assembling the best team will maximize your chances of achieving the project you envisioned, on time and within budget. Choose the wrong candidates and you can bet that disaster awaits.

With so much at stake, how do you decide? The construction industry abounds in choices of building professionals, including architects, designers, builders and general contractors who work in different ways. *New House/More House* will introduce you to each player and describe what responsibilities each has and how each accomplishes his tasks. You can actually team-up professionals in different combinations to produce the same project.

The first variable in this selection process is that homeowners have differing circumstances. Some have the time to carefully plan their ultimate dream home and have no deadline to meet. Others must have space as soon as possible. You may have just discovered twins are due in seven months and extra bedrooms have to be ready. A few lucky homeowners have unlimited resources and envision large houses with cutting-edge design. The majority have limited budgets and must maximize what their building dollars buy.

Your own circumstances will influence the choice of a project team. You are alone in deciding to whom you will entrust your project, which contains so much of your emotion, time and money. Your job is to identify potential candidates and match their expertise with your goals and requirements. For example, if space is needed quickly, hire someone who is known to finish projects on time but isn't into constructing mansions. If the latest architectural style shown in the high-end magazines is your goal, then you will no doubt select a prominent designer who may not work particularly at a fast pace. Planning a gourmet kitchen remodeling? Don't hire a company that specializes in constructing garages. Do these choices seem obvious? They sure do, but you would be surprised by the choices some homeowners make in selecting their building team.

Locating, qualifying, and hiring the right professionals for your project is a process of asking the right questions, understanding each professional's role, and doing the proper background research. Selecting the right architect or contractor means more than deciding on a flashy designer or the carpenter who still cuts lumber by hand. You should choose building professionals only after evaluating them carefully on the basis of experience, communication skills, teamwork, and references.

Potential candidates should be able to demonstrate their abilities through past projects. At this point, you have to do some legwork. You can learn how to interview prospective professionals to determine the types of services they offer and how they differ. *New House/More House* will supply you with the right questions to ask and the answers you want to hear. As I mentioned in the introduction, choosing the right people for a building project is a lot like buying a car. You first go to the showroom, find a car that catches your eye and circle it a few times, kicking the tires for good measure. Then you jump in and take a test drive to get the feel. You check out the colors, accessories, and warranty program. But before you plunk down your hard-earned cash, you run to the library to double-check the repair history of previous models. In other words, you do your homework.

Qualifying professionals also includes evaluating the legal agreements they utilize. Proper contracts are a must to protect your rights throughout the project. Due to our current legal climate where litigation is commonplace for errors or injury, a good contract could spell the difference between a quick resolution or a protracted disagreement.

Just as building professionals come in all shapes and sizes, so, too, do their legal agreements. They range from a dollar amount scrawled on the back of a business card to a twenty-page contract with fine print carefully written by a clever attorney. Part of the reason for interviewing potential professionals is to assess as early as possible each candidate's contract. Although you may have found the best builder for your project, his contract may put you at such a disadvantage that you must reject his services.

New House/More House will introduce you to several different contract types. A good agreement should not be weighted in anyone's favor, but should be fair and equitable to all parties. I will try to breathe some life into a dull subject by showing you key clauses that will safeguard your interests as well as pointing out the traps you should avoid. Talking about a bunch of legal mumbo jumbo can be about as exciting as watching grass grow!

These topics relating to hiring the right professionals are covered in chapters 2 through 5, and 7.

Step Three: Have the Project Designed to Meet Your Expectations

We have mentioned that this whole process starts with you identifying a need to change your living environment. Depending on their abilities, some homeowners have no problem envisioning how their project will appear, while many others will have no clue. Some may be adept at verbalizing their goals, while many are either intimidated or indecisive in dealing with these issues. Wherever you fall in the above categories, at some point you will be presented with an architectural design that you must either approve or send back for revision.

The third step of our program is to ensure that the final design matches or exceeds your expectations. Effective communication between you and the design professional is the only way to maximize your results. As an architect, I know my job is to understand your ideas regardless of their merit and extract as much of your input as I can. I then have to translate these ideas into a three-dimensional reality that you can understand. The worst comment an architect or builder can hear from a client walking through a newly built home is, "I didn't think it would look like this. I hate it!"

New House/More House will teach you how to focus your visions and goals by preparing a written statement before you begin the hiring process. We will show you how to establish your primary and secondary requirements. I will encourage you to start with as many ideas as pop into your head and then prioritize them from top to bottom. This written "program" or statement is also a useful tool to communicate your project's scope to potential building professionals. We are not looking for a literary masterpiece, just an informal program that lists the criteria for your project.

The program can be in any format that makes you feel comfortable expressing yourself. You will probably start off talking about the functional rooms you require, explaining what you want that space to accomplish. You can add more specific detail to paint in any frills you desire, anything from a hot tub for six to a kitchen worthy of a gourmet. Your program should also contain the amount of money you expect to spend for the project and a timetable for starting and finishing.

At this stage of the game, when you are just beginning to visualize a project, I recommend reading books, magazines and newspapers with lots of pictures. You should start organizing a file by clipping pictures of anything that appeals to you, from design styles to a hot tub. The Internet is a wonderful resource to locate all kinds of products and information. If you have difficulty visualizing or expressing your ideas for your potential project, your file can go a long way to help you shape project ideas and communicate effectively.

As you begin working on the design with a professional, you must be able to understand what the designer is proposing. Most people haven't learned to read architectural drawings. Traditional construction drawings take a three-dimensional house and only illustrate it in two dimensions. Everything that has depth appears flat on the page! Taking on a homebuilding project doesn't mean you need to run to night school to take a blueprint reading course. Instead, *New House/More House* will show how you can work with your design professional in three dimensions.

Architectural models, hand drawn perspective sketches, and computer-generated, three-dimensional modeling will dramatically increase your comprehension of the proposed design. Instead of timidly accepting design solutions shown on drawings you can't understand, these aids will make you an active participant.

The final goal in this step goes beyond the design phase. Every design decision you make must be ac-

curately transmitted to the construction team. Your design professional must take the preliminary design and create very detailed construction drawings and specifications that account for every element in the project.

Here is yet another point where disaster often rears its ugly head. Despite all your work to carefully select every material and color, the information may fail to appear on the drawings, and the contractor will not included it in his bid. Suppose you change your mind several times about the color of that hot tub, but the final drawings reflect your first choice instead of the last. Taking out the wrong tub and installing the new one could create a cost overrun.

Your job as executive project manager is to verify that all your design selections are included in the drawings and construction bids. Armed with checklists from *New House/More House*, you will be able to ask the right questions to double check all these items. Does this sound like a lot of work? Maybe, but most projects can be reviewed effectively in one or two meetings. After all, you don't want to live with a daily reminder that somebody messed up!

Chapters 6, 8 and 9 are devoted to helping you realize the project you expect.

Step Four: Control the Cost

We've already mentioned the subject of cost overruns several times in previous steps. Paying more money than you planned usually ruins your day. We've all heard countless stories about how a budget figure, agreed upon at the start of a project, seems to grow like a weed as the project progresses! Should you prepare yourself to fight those guys you hired to build your project? Not so fast! Budget busting is often caused by the designer or contractor, but the biggest culprit in my experience is usually the homeowner. Here are two examples:

You arrive at the design stage of a project and commit to a project scope, size and budget. Yet as every week goes by, you add one "little" change, like marble instead of carpet in the front entry, or solid wood doors instead of the less expensive hollow hardboard type. How much could this "one little change" cost? Not much you reason. Wrong! These weekly changes can add up fast and give you a nasty surprise when the bottom line is totaled.

Or how about the time during construction when you decide to bring in an interior designer to help with some furnishing ideas? She shows you an oversized bed you can't live without. Oops! A door is now in the wrong location because of the extra wall space needed for the larger bed. "Just move the door," you tell the contractor, who in turn authorizes the carpenter to spend most of a day re-framing a wall and calls in the electrician to move an outlet. One thing you will find out quickly is that in today's busy world, nobody throws in construction changes for free. Time is money. If there is additional work, you will pay extra. *New House/More House* will explain how you can avoid these expensive traps. Remember, it is far cheaper to erase the door location on a drawing than to rebuild it on the site! I will also show you how to avoid expanding your design program and minimize "budget creep."

To keep costs down, I stress establishing competition in fees and services from your building professionals. Let's return to our car analogy. As a smart consumer, when you finally decide on a make and model, do you buy it from the first dealer who gives a price? Unless the salesman is your uncle, you would probably visit other dealers and negotiate additional prices. Come to think of it, depending on your uncle, you might want to check out other deals anyway!

If I'm asked to submit a proposal for architectural or construction services and I know there will be others competing, I must "sharpen my pencil" in order to get the project. Conversely, if I am the only bidder, then the price will probably be higher. Com-

petition is the easiest way to get "the biggest bang for the buck." *New House/More House* will show you how to introduce competition into the process and ensure that everyone is providing the same services.

Controlling the cost would not be complete without discussing financing your project. For most of us, a building project is the biggest investment and financial transaction of a lifetime. *New House/More House* will discuss financial tools such as construction loans, bridge loans, and end loans. I will point out how building professionals are paid and at what point funds will be required. Today's competitive money markets have opened up more choices than ever for homeowners to finance a project. Here again, competition and service reigns supreme. Controlling lending costs can save you as much as keeping the design program in check.

Controlling project costs are discussed in detail in chapters 9 and 10.

Step Five: Control the Quality

The fifth step in the home building process addresses controlling the quality throughout design and construction. How do you know if the materials specified by the architect will be long lasting? How can you tell whether the contractor is using proper methods to build the project? Unfortunately, the vast majority of homeowners are not equipped to analyze quality. Most assume that the local building inspector from city hall will be on the job to safeguard their interests. Building departments and the building permit process actually cover only a portion of a construction project.

As we have just mentioned, building departments are interested in protecting the health and safety of the public. They focus on sanitary issues like plumbing piping that will deliver safe drinking water and electrical connections that won't explode when you plug in a toaster. Fire departments are interested in warning devices that will alert you to fire and smoke. But when it comes to cosmetic finishes like tight carpentry trim joints, or ceramic tile installed in straight lines, you could be on your own.

Quality is a very subjective topic. What may well be acceptable workmanship to one tradesman is unacceptable to another. Robert Pirsig's *Zen and the Art of Motorcycle Maintenance*, a popular book out in the 1970s, devoted itself to defining what quality really is. The author ended up in a mental hospital because the subject is so impossible to define. You must decide your own level of design and construction quality and be prepared to monitor your project for its presence.

The homeowner has three choices to assess quality. Hiring competent building professionals, as we mentioned before, is by far the best. Some professionals spend a lifetime building their reputation by delivering quality work. Others just want to make their money and move on to the next project. Good tradesmen are in the best position to judge their own quality because they are on the job every day and assemble every piece. These are the guys you want to find.

Your second choice could be hiring an outside consultant to review the plans or your own architect to visit the construction site. This is an expensive alternative and may be needed only when problems have already cropped up. If things are going from bad to worse on your project, it may well be cheap insurance to bring in a hired gun to check the scene.

You, the homeowner, are the last and most common alternative. Most people can look at a door and see if it's square and swings true. But can you look at newly installed heating ductwork and decide if it has been sized correctly? Do you know at what locations plumbing piping should be secured to the wall? I doubt you do. Nor should you have to.

New House/More House presents a program which will enable you to rely on the expertise of others,

and yet retain managerial control of your building project. Your ultimate weapon is the purse strings. It's your money and only you can authorize funds to be released or paid. Money means leverage! Building professionals occasionally must be treated like children. If they are good, they receive their allowance. Bad behavior sentences them to sit in the corner until they are ready to act properly. Once you have paid for work done, it is far more difficult to fix existing problems than if you had not written the check. However, if you withhold all or too much payment, you may find no one returning to work on your project.

Can you tell from this brief discussion that controlling quality is a sticky issue? Your best safeguard is to hire reputable professionals. Our program stresses that reputation counts for nearly everything in the residential construction industry.

Chapter 11 will guide you through several strategies to control quality throughout your project.

Step Six: Warrant the Project

We have reached our last step, which signals light at the end of the tunnel! At this stage construction has finally finished and you have moved in. Now you get to enjoy the project you and your building team created. Even the best level of construction workmanship can experience conditions requiring adjustment or repairs. *New House/More House* recommends a comprehensive warranty program be in place to address post construction problems. Regardless of the size, from big to small, all projects need warranty protection.

Construction warranties come in all sizes and shapes and are not created equal. The best warranty you can get is simple; all materials and workmanship are warranted for a specific period of time. Unfortunately, many warranties have exclusions and restrictions which limit their effectiveness. For example, one contractor's warranty may state "Concrete and drywall are construction products where cracking is unavoidable due to the nature of the material."

I will explain warranties in detail, pointing out what should be included as well as indicating the clauses that some builders favor, conveniently limiting their responsibility. In addition to reviewing a building professional's contract as part of the interviewing process, you should ask to see a copy of their warranty. If you are presented with a lot of fine print drafted by an attorney, chances are your rights will be limited. This alone is cause to reject the prospective candidate. Some people who trustfully sign unread contracts may have actually removed their warranty rights normally received under prevailing state law.

Generally, you have one year to identify any problem and obtain repair under the warranty. I will pass along tips for getting the most out of your warranty program. Many homeowners also don't realize that many materials, such as roofing, siding and appliances, carry separate, longer manufacturers' guarantees.

Extended warranty programs from national insurance companies are also available to supplement your builder's warranty. *New House/More House* will introduce these policies to you, explaining the pros and cons. Careful review of these multi-page documents is required; often they have as many holes as Swiss cheese! You must weigh the additional cost of these programs versus the value of extra protection provided.

Builders and contractors' services vary during the warranty process. Some will come promptly to your house to investigate and repair problems. Others will procrastinate and will require many calls and threats before they show up. This is yet another facet of the need to qualify prospective building professionals before hiring. When you call their references, be sure to check out their track record during the warranty period. As always, we consistently return to the subject of a professional's reputation.

Chapter 12 is dedicated to discussing warranty topics in detail.

Reviewing our six steps should have alerted you to a singular fact: Armed with the necessary knowledge, you can maximize your outcome if you are willing to go the extra yard. As in most other endeavors, the more effort you put forth, the better the outcome.

Chapter One Recap

The Six-Step Program for Project Success

Step One: Learn the Process

Residential projects are divided into three phases.

- *Predesign* identifies the project's scope.
- *Design Phase* produces an architectural design that fulfills your goals and requirements.
- *Construction Phase* builds the project, pays the construction professionals and provides a warranty.

Step Two: Hire the Right Professionals

- Perhaps the most important step, potential architects, general contractors and builders are evaluated by past experience and references to find the best candidates for your project.

Step Three: Have the Project Designed to Meet Your Expectations

- Homeowners must focus their project requirements and effectively communicate this information to their building professionals. In return, building professionals must help their clients clearly understand the proposed design, using drawings, sketches, and three-dimensional models.

Step Four: Control the Cost

- Cost overruns are one of the most common problems experienced on a project, and can be caused by several factors. Architects could specify the wrong product or general contractors and builders may forget to include all project components in their pricing.
- Homeowners are also frequently to blame due to continually expanding their design requirements or making changes during construction.
- Competition in design fees, construction costs and financing fees are one of the best ways to win the budget battle.

Step Five: Control the Quality

- Average homeowners may not be equipped to judge construction workmanship quality for themselves. Outside assistance could be beneficial to safeguard your interests.
- Only paying for acceptable workmanship is an effective method of controlling the quality on your project.

Step Six: Warrant the Project

- A comprehensive warranty program is a must for addressing any post-construction problems that may appear.
- Warranties are not created equally. Many have exclusions which may leave you with little protection.

Chapter 2

The Building Team: What It Does and How It Works

After the conclusion of a seminar I had just presented to a group of homeowners, several participants shared their homebuilding experiences with me. One man complained that his architect and general contractor never agreed. "Last week they stopped speaking to each other over a dispute on the location of installing the heating ductwork. The contractor wanted to hang the ducts under the ceiling and enclose them with drywall. The architect practically had a heart attack, claiming the line of his ceiling design would be destroyed. Now I am stuck in the middle trying to get them to make up and finish my house!"

A woman also told her story. "I had several problems called to my attention during the construction of my addition. With each problem, the architect and builder asked for more money to cover conditions they hadn't expected. But my architect and builder were on the best of terms and never disagreed. I'm not sure who works for whom, but I don't think the architect was working directly for me."

As we discussed several times in the Six-Step Program, hiring the proper building team to construct a new house, addition or remodeling is your most critical decision in producing a successful project. Select the right group and you have a great chance to enjoy the entire process. Choose the wrong players and your project can end up like a typical season for my favorite baseball team, the Chicago Cubs: "Wait 'til next year!"

The two stories related to me are good examples of homeowners who did not fully understand the importance of working relationships within the building team. After I asked a few questions of the man, he admitted that he discovered after construction started that his architect and general contractor were involved in a lawsuit years ago and used every opportunity to seek revenge. "If I had known this earlier, I would never had paired them up," he explained.

I didn't even need to ask any further questions about the second story. It was obvious the woman had hired a builder who was responsible for providing the architectural design, as well as constructing the project. The architect always agreed with the builder because he signed the architect's paychecks!

It is very important for you to understand the role each building professional has in the process and how they are combined to produce the project. Too often homeowners hire a building team without understanding each professional's role. They have a vague idea that someone provides the plans for several tradesmen to construct, yet they don't know the difference between contractors, carpenters, and cabinetmakers. Each has a very specific role in today's specialized construction world, with different backgrounds, training, and experience. The method of constructing a project has changed over the years; now every trade is a specialty unto itself.

In my early days, as a young architect, a grizzled carpenter chose to impart his grudge to the new kid on the block. "In the old days we used to do everything, pour the concrete, nail up the wall framing, install the shingles, hang the doors and even lay the tile. I was on the job for the duration. Now there is a separate crew for everything. Today I finish the rough carpentry framing and move on to the next job. I never get to see the finished product." His story is all too true. Constructing even a simple building project will require fifteen to twenty different specialty trades coming and going from the site. We can divide all the players in the industry into design professionals and construction specialists. Learning who coordinates each participant's work, is therefore very useful. At this point, the following descriptions are general in nature as we will be exploring their activities in greater depth in later chapters.

Design Professionals

This category includes architects, architectural designers, and interior designers. They have the responsibility of assessing the project requirements, designing a solution and graphically representing these ideas in a language depicted on drawings. Besides these three design professionals, specialty designers for kitchens, baths and lighting are also available, usually working through retail stores.

Architect and Architectural Designer

Do you have the image of an architect working over a drawing board with a T square and pencil? Those days are fast disappearing as architects are now seated in front of a computer instead of drawing by hand. Although the technology has changed, the tasks are still the same. The architect is responsible for designing the project in a language of descriptive terms and recognizable images both the homeowner and builder can understand. Architects are really working on two levels. Determining the design is a process of understanding their client's project requirements and then developing an artistic yet functional solution. The approved design is subsequently translated by the architect into a technical language shown on drawings that will be used to construct the project.

Consequently, the architect must be well versed in both the abstract ideas of design as well as the nuts and bolts of construction. Using both sides of the brain is a must. Unfortunately, not all architects are on the same plane. Some are better designers; others are more technically oriented. The best architects can do both equally well.

Today's architects receive their training at university architectural schools. Their course of study require both liberal arts and technical classes, providing a well-rounded, four- to six-year study of architecture. Once they receive their degrees, they train as employed interns under the supervision of a licensed architect. After accumulating several years of working experience, they must demonstrate their capabilities by passing a rigorous national exam.

Practitioners of architecture are not allowed to call themselves architects unless they are licensed. Having a license has become increasingly important as many building departments, especially in urban and suburban areas, will not issue a construction permit unless the plans have been sealed by a licensed architect. The American Institute of Architects (AIA), of which I am a member, is a national organization of licensed architects and promotes good standards of practice.

In addition to being a designer who is knowledgeable about construction, the architect should be a good communicator, listener, and analyst. Good design is actually the product of architect and homeowner working together to accomplish a common goal. The better they work together, the better the result. Architects must be perceptive in learning their client's requirements. Often this information gathering process can be as difficult as extracting an impacted tooth! Many homeowners have a problem focusing and verbalizing their wishes. If the project's scope is unclear, the architect should inject discipline into the process, organizing and guiding their clients on a logical path.

Although architects have similar educational backgrounds, they do not necessarily practice their profession in similar ways. Residential architects come in two varieties: those who work directly for the homeowner on custom projects, and those who work for builders or developers. While the architect working directly for the homeowner has continuous contact with his client, the production architect hired by a builder may have limited or no contact at all with the homeowner. Recognizing this difference is very important in selecting the right architect.

Architectural designers are practitioners of architecture who are not licensed. They can vary in expertise from a draftsman who can prepare basic drawings to an individual with an architectural degree who never took or passed the licensing exam. Drawing a parallel between an architectural designer and a licensed architect is comparable to a nurse who has quite a bit of medical training and experience, but has not reached the level of a doctor.

Architectural designers can work in many areas of the country where building departments do not require a licensed architect's seal on drawings submitted for building permit. A simple call to your local department can verify whether this is a requirement. Since designers are not licensed, the homeowner must verify each designer's knowledge and experience. Their ability to adequately perform is dependent on their on-the-job training on certain types of projects. On straightforward assignments,

an architectural designer may be an appropriate choice. Projects involving a higher level of design or technical complexity usually are better served by a licensed architect.

Interior Designer

Traditionally, interior designers known as "decorators" selected nonpermanent materials such as furniture, wallpaper, window coverings, flooring, and color schemes. During the last ten to fifteen years, interior design has evolved into a profession. Many practitioners today receive advanced training at interior design schools, and in some states are licensed. The American Society of Interior Designers (ASID) is a national organization that promotes the profession.

Besides their traditional scope of work, interior designers now are commonly involved with plan layouts, custom cabinetry, kitchen and bath design and accent lighting. Their input during the design process can be very beneficial in developing the fullest potential of a project. Choosing to employ both an architect and interior designer on your project will require some coordination on your part. This increased scope of interior design service has also created opportunities for conflict in design philosophies. Architects and interior designers often view project design in different ways. They may not always agree on a common course of action. As the old maxim reminds us, a camel is a horse that was designed by committee! You must carefully understand the exact services each design professional will provide. Your goal is to eliminate the possibility of paying twice for overlapping design work, or worse, design work assumed to be the other's responsibility that is overlooked by both professionals. *New House/More House* will discuss the proper way to integrate these two design entities.

Construction Professionals

The second category of team players consists of the group who construct the project. I divide this set into four separate categories: general contractor, subcontractor, builder, and developer. They have the responsibility of taking the design plans, assessing the cost of construction, and assembling the team of specialty tradesmen to schedule delivery of materials and build the project.

General Contractor

A general contractor is responsible for constructing the project from architectural drawings provided by the homeowner's design professional. He does not typically enter the project process until the plans are nearly finished and the majority of design decisions are complete. The term "general" in general contractor means he is in charge of the specialty subcontractors that construct the project. The "general" hold the "contracts" with each of these individual trades, thus the title "general contractor."

General contractors, or "generals" for short, come in all different sizes and shapes. They may work alone from the back of a pickup truck or work for large companies with many employees. Their training ranges from learning on the job to earning university degrees in construction programs. Working for the homeowner, the general contractor first determines the scope of the project from the architectural drawings, sizing up which materials have been specified and which tradesmen will be required to construct those particular items. For example, if the plans call for a brick fireplace, a mason must be employed. Wood flooring requires hiring a company that specializes in installing this particular product.

Constructing a new house or large addition requires over twenty different specialty trades. A typical remodeling project could require more than ten. The electrician, plumber, and carpenter on your project do not work for you directly; they are hired and paid by the general contractor. The general determines the project schedule and coordinates the sequence in which materials and tradesmen arrive at the site. As construction progresses, the general supervises the accuracy and quality of construction. Ultimately,

the specialty trades are responsible to the general contractor and in turn, the general is responsible to you. Homeowners frequently make a mess of things by trying to direct the tradesmen on the construction site. Actually, they take orders only from the con-ractor. Many general contractors have their own carpenters on staff, but the majority of trades are independently hired.

A general contractor must be knowledgeable about construction practices and familiar with local building codes. Years of experience are required to learn the intricacies of each trade and the construction process as a whole. Typically, many generals start out in the trades, graduating from carpenter or electrician to a more managerial job. On small projects, some contractors wear two hats; they may serve as a working carpenter as well as handle the contractor's chores.

Managerial skills are essential to effectively coordinate materials, deliveries, and day-to-day labor schedules. Business acumen is also a priority, considering the large amounts of money that change hands during a project and the number of legal agreements that will be negotiated and signed. Like the architect, the general should possess good oral and written communication abilities. A good contractor spends half his day at the job site and the other half on a cell phone. Construction requires a huge amount of verbal instruction, including questions and answers.

Regulation of the qualifications of general contractors can vary widely from location to location. Some states require generals to be licensed through examination. They are thoroughly tested to determine their knowledge of building codes, construction practices, and contract responsibilities. Licensing through examination insures a higher level of competency in construction practice. Other states have no requirements; anyone wanting to label themselves as a general contractor is free to do so. Some contractors who claim to be licensed may have actually only paid a registration fee to the state! This is not a license based on knowledge, but a means for state and local governments to collect revenue. Licensed contractors as well as architects have more at stake in conducting their daily activities than their unlicensed brethren. Misconduct can be a cause for losing their licenses, and thus their livelihoods. Unlicensed practitioners have no such limitations.

Subcontractor

A subcontractor is contractually employed by a builder or general contractor to provide specific construction trades. These are the guys with the tools. They have several nicknames around the construction site: electrician (sparky), mason (brickee), heating (tin knocker), carpenter (wood butcher) and plumber (Mr. Plumber). Subcontractors, known as "subs," are at the heart of the construction process. Each sub that is working for a contractor or builder is actually a separate company with its own set of owners and workers. It is not unusual on a new house project to have more than twenty separate companies involved, each looking out for their own interests and bottom lines.

Each sub is also responsible for ordering and supplying his own construction materials. For example, the carpenter supplies the lumber, the electrician supplies the lights, and the plumber furnishes the bathroom fixtures. That adds another twenty supply companies to the mix, for a total of more than forty separate business entities working together on a project. You may now begin to see why the chances for errors and delays are magnified on a construction project.

A builder or general contractor's reputation sinks or swims with the reliability and expertise of their subcontractors. If a sub's work is unacceptable or he experiences financial difficulties, such as not paying his suppliers, the general must step in to correct the situation and preserve his own reputation. Many a horror story occurs when a builder, trying to maximize profit, tries out a new less-expensive subcontractor only to find that the sub's poor performance wasn't worth the lower price.

Therefore, a mutually beneficial relationship exists between general contractors and their subs. Generals and builders prefer to use the same set of subs on every new project. They like the peace of mind that a continuing business relationship provides. Knowing that you can count on a particular sub who furnishes good workmanship, shows up on time, and correctly estimates the work makes life for the general a lot easier. In return for the contractor feeding his subcontractors a steady stream of work, the subs in turn take care of their client with the best of service at fair prices. The old saying, You scratch my back, I'll scratch yours, certainly applies to this relationship. In short, the better the subs, the better the product.

Many states and local governments require subcontractors to be licensed or bonded. Electricians and plumbers, for example, may be required to display appropriate training and experience through an apprentice program or by passing an examination. This ensures a level of competence for trades whose work can affect public safety and health. Bonding is a form of insurance that underwrites specific construction work that is not installed correctly, again for the protection of the public. If a subcontractor's work does not meet legal requirements, for example, failing to install plumbing that meets sanitary standards, the bonding company will pay to correct the problem if the plumber is unwilling to correct the faulty work. It is very rare that a bonding company is required to pay for such repairs. Just as you value your driver's license, licensed tradesmen also need to maintain their ability to work.

Builder

Homeowners frequently confuse the roles general contractors and builders play in the project process. The builder is responsible to the homeowner for providing the entire project, including architectural design and construction. Unlike the general, who enters the process after the architectural work has been completed, the builder becomes involved at the start of the project. The key difference between these two construction professionals is this additional dimension provided by the builder in furnishing architectural design and controlling the project's cost. In the past, before the days of outsourcing, large builders had architects and designers working directly on their own staff. Today the builder usually hires an independent architectural firm to furnish the design and plans.

Once a builder completes the planning stage of a project, he proceeds to construct the job in the same manner as a general contractor. Some builders were formerly general contractors but preferred the benefits of controlling the architectural design. On the other hand, generals often want nothing to do with the design process. "Just give me the plans and get out of my way" is their philosophy! Occasionally builders will work as general contractors if a predesigned project appears to be a good business deal, and generals have been known to bring in an architect if it means getting a lucrative project. For the most part though, each professional sticks to their traditional role.

Builders may also become involved in providing the land site for new construction. They often are offered new residential subdivision lots for sale in advance of the public, gobbling up the choice lots at early bird prices. This is beneficial for the builder; not only does he make a profit reselling the land, but he can charge a premium for constructing the project. If you fall in love with a particular lot owned by a builder, you are stuck working with them to design and build the house.

Developer

Developers are much like builders, but take the process one step further. Ranging in size from local to national companies, their first step is to find undeveloped raw land suitable for new home sites. Unlike the builder, who starts on a site ready for con-

struction, the developer plans the roads, lot sizes and utilities to service the development.

Today's governmental control over land development presents a series of hurdles for the developer, often requiring a wait of months or years before the first shovelful of dirt can be turned. Development has an impact on the road system as well as municipal water, waste treatment, and utility networks. New residential neighborhoods also impact school systems, police and fire protection, all adding expense to a local municipality's budget. Lately, many governmental jurisdictions require developers to pay impact fees to cover these expenses. These are figured on a per lot basis and the developer passes along these additional costs to the homeowner.

After the developer divides the land into buildable lots and obtains government approval, different predesigned houses are offered for sale. Several display models are constructed from these plans for marketing purposes. Specific houses usually will not be built until a sales contract has been signed by the homeowner, stipulating the chosen model and optional features. At this point, the developer changes hats and acts like a builder, constructing the house with his own forces and other subcontractors. Customizing the design or making changes to the plans requested by the homeowner is usually very limited. Homeowners have only the choice of available finish upgrades and color selection.

Developers make their money by repetitively building the same basic house models. They have great leverage in buying labor and materials at steep discounts. I remember one developer I worked with who had the following practice. After his subcontractors had built two or three of the same house model, he would tell them, "Now that you know how to build this model, you should be able to trim down your construction time. Therefore, I want you to cut your price by 15 percent on all future models!" You can see that developers have construction pricing down to a science.

Since *New House/More House* deals predominately with custom projects, only a few of our strategies may apply to subdivision homes. I will point out these areas as we approach each discussion.

The Final Player

I almost forgot to include the most important team member—you, the manager! Homeowners come in all sizes and shapes and with varying degrees of knowledge. Some feel they know everything, and will not listen to the good advice provided by the professionals they hire. The smart ones admit they know next-to-nothing and are willing to learn, applying their best judgement along the way. Their involvement and decisions can make or break a project.

Project Delivery Systems

Now that you have become acquainted with building professionals, it's time to learn how they work together. As I have mentioned before, hiring the right professional is the homeowner's most critical decision. You have two basic choices in selecting how these professionals provide their services. The first method I call the *Traditional System*, and the second is the *Design/Build System*. Figure 2-1, *Project Delivery Systems*, on page 22, illustrates the flow of responsibility between professionals and the homeowner.

Although the professionals perform many of the same tasks in both options, they are contractually combined in different ways. You can observe from the diagram that each option has certain professionals working directly for you, while others are working directly for someone else. These contractual relationships reflect who will be loyal and legally responsible to you. Each system has certain advantages and disadvantages depending on your circumstances. The bottom line is that whichever option you choose, you still get to enjoy the same product at the end of the process. Let's examine the advantages and disadvantages of each system.

Traditional System

Using the Traditional System, you create a team by separately hiring an *architect* and a *general contractor*. I call this system "Traditional" as it has been the most common method of accomplishing a project. The Traditional System is used for custom new homes, additions and remodelings. It can be an effective means for delivering a project if all parties communicate and function well as a team. The following discussion illustrates the advantages and disadvantages for this system. As indicated in the figure, you have two distinct contractual relationships.

Advantages

• Choice is the name of the game here. You determine the best architect and general contractor for your project and negotiate their fees and services. For example, if your project calls for a particular design style, the Traditional System allows you the freedom to identify architects with that specific expertise. Developing a comfortable relationship with your design professional is all important in getting the project you envisioned. Higher design expectations require more architectural design mastery. If you intend to build a Taj Mahal, you better find a designer who has a lot of palace design experience!

• Having an architect working directly for you means you have an agent safeguarding your interests during the project. He can offer continuing guidance throughout the process, hopefully steering you around obstacles and clear of pitfalls. A full-service architect can provide many different professional services starting with analyzing potential building sites before you buy. After the plans have been completed, the architect can help you identify general contractor candidates, advising you on their qualifications and assisting in the interviewing process. Would you feel uncomfortable asking leading questions during a contractor interview, such as, "Do you have the financial resources to complete this project?" or "I have heard many good things about

Figure 2-1

Project Delivery Systems

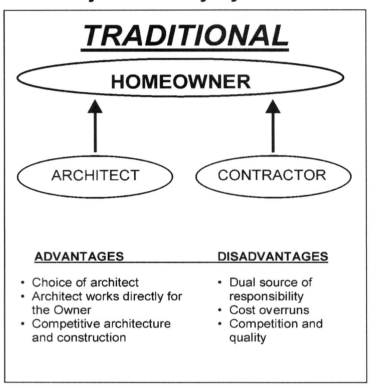

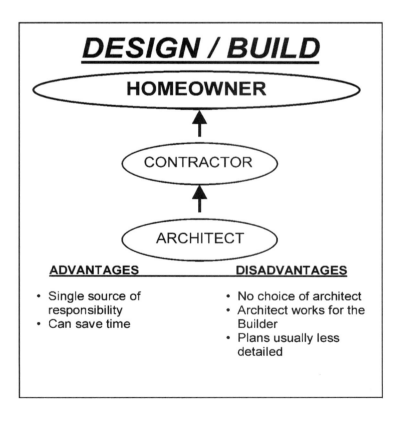

your work, except for that one house you built down the street where the chimney fell down." As an architect, I have often asked these difficult questions for my client.

Negotiating construction contracts is another good service provided by architects. Let's assume you are ready to award the construction contract for your project, and you have two qualified contractors who are a few thousand dollars apart, but still over budget. You would love to have them lower their bids just a bit, but you're too embarrassed to ask. Your architect can jump in, telling the two bidders that if they want the project, they better knock-off some more money, or the other guy gets the job!

During construction, the architect can visit the site and advise you if the contractor is building in accordance with the drawings and if you are receiving quality workmanship. If you are unfamiliar with construction, having your own architect is advantageous, as you are not alone in dealing with the contractor.

• Competition is assured throughout the project. If you follow our program's suggestions about defining your project goals and requirements prior to hiring a design professional, you are in a position to solicit competitive fees from several architects. With a thoroughly prepared set of project plans, you can also benefit from competitive bidding by a group of qualified general contractors. I know of no better means to stretch your project dollars than by encouraging competition on all levels. Remember our analogy about purchasing a car? Don't buy from the first guy you talk with; check out the options, play the field! That's the way to maximize your value.

Disadvantages

• You have just hired two separate building professionals, pairing two independent companies and yourself to accomplish the project. Either you have created a smooth-running three-member team, or

you have given birth to a three-headed monster! Many a catastrophe has occurred when your architect and general contractor do not work well together. As the owner, you are caught in the middle having to resolve disagreements and conflicts.

For example, some architects are rather sensitive about their design work. It can be compared to giving birth to a new child. If the contractor decides to take a few "liberties" with building the design, the architect can go ballistic about messing with his baby! On the other hand, if your architect can't stop redesigning the project after construction has begun, the contractor will sooner or later tell you this guy is out of control and that construction is stopping today unless you do something!

Kidding aside, the problems tend to be more specific. A typical dispute may develop as follows: the architect fails to specify on the drawings the manufacturer and model number of the doorknobs you selected. The general contractor can do one of three things during the bidding process. He could call and get a clarification from the architect, or instead, he could include an allowance of so many dollars for each knob within his bid. The worst case scenario would find him excluding door hardware altogether since it wasn't included in the plans. Perhaps he thought you were going to supply your own antique knobs from grandma's house!

When the day comes at the construction site for the hardware to be installed, the contractor installs doorknobs you never saw before. You ask, "Where is the hardware I chose?" The contractor replies that the plans were unclear and this is what he included in his bid. Now you call the architect for an explanation and his response is that the contractor should have asked for more specific information during the bidding phase.

The architect and contractor each insist their position is correct and look to you for direction. You are now in the classic dilemma of deciding who is right and who is at fault. Our typical dispute could be resolved in a number of ways, but in my experience, homeowners are usually in the middle holding the bag, reaching in their own pocket for more money. *New House/More House* will later address in detail several methods to minimize this "triangle of conflict."

• Another cause for disaster stories springs from receiving construction bids above the architect's estimate. Architects are notorious for underestimating construction costs. Many use simple rule-of-thumb methods to determine construction cost. Also, they may not be conscientious about updating budgets as changes or additions are made during the design process. One of the worst surprises you can experience after the completion of design work is to discover that your contractor bids are above your budget. To continue the project, you must either come up with more money, redesign to reduce the cost, or a combination of both. Either choice is no fun; your expectations must be lowered and your schedule may be delayed if redrawing is required. Then to add insult to injury, your architect submits a bill for additional design work! Sound pretty dismal? We will discuss strategies to avoid these potential money traps.

• Finally, let's look at competition one more time. While I earlier touted the fact that competitive bidding produces the lowest project cost, it may not always guarantee the best quality. You can always find a cheaper price, but is it worth lousy service and poor workmanship? You must determine the quality and attention you expect at the beginning of a project. Prior to gathering proposals from architects and bids from contractors, you must spend time examining and researching their reputations and experience. I call this process *"pre-qualification."* Chapter 3 advises you not to invite any professionals to submit a proposal unless your research shows that their work fits your qualification parameters and that you feel comfortable with them.

Design/Build System

Our second project delivery option is the Design/Build System. From figure 2-1, you will observe that the homeowner now has a single source of responsibility to design and construct their project—the builder. Unlike the Traditional System, the architect works for the builder, not the homeowner. Although still a very important player, the architect is truly just another subcontractor to the builder, like a carpenter or electrician.

The Design/Build System is utilized almost exclusively for producing new tract homes in large subdivisions and semicustom homes in other developments. The builder/developer determines the architectural design for an unknown future homeowner. This system can also be an effective method for building custom homes, additions and remodelings. Let's discuss the pros and cons of this option.

Advantages

• The first major advantage is that the "triangle of conflict," identified as a disadvantage in the Traditional System, disappears. Since the builder is solely responsible for the project, you only have one company to deal with. If the architect makes a mistake, the error is the builder's problem, not yours. Just as the builder, at his own expense, would have to change a wall the carpenter built in the wrong location, so too must he cover architectural foul-ups.

Obviously, this relationship can eliminate many sources of conflict. Your builder and his architect will be the best of pals, since the architect knows who signs his paycheck. You will never see any finger-pointing over-cost extras between these two.

• Cost control and preliminary budget accuracy is enhanced in the Design/Build System because the builder is on board from the very start. As the architect develops the design, the builder, who is the construction cost expert, is generating accurate estimates. The uncertainty of project cost is greatly reduced since the architect is no longer the primary source for preliminary construction "guesstimates."

If the project's scope increases, the builder can quickly tell you an accurate cost for the additional space. Or if you are prone to add changes at every design meeting, the builder can keep a running tab. This affords you the comfort of always knowing the project's cost, eliminating nasty surprises down the road and saving redesign time. Once the design work is complete, a guaranteed cost is already in place.

• Saving time is the third benefit of this system. Since the builder is busily preparing cost estimates during design, the time required for bidding by general contractors in the Traditional System is essentially eliminated. You also save time by not having to identify, prequalify and interview potential general contractor candidates, let alone negotiate another contract. You selected your builder up front, so this cumbersome chore is removed from the process. These two elements can reduce the time length of a project by as much as one or two months, depending on the project's size. Even the building permit application period can be conducted concurrently with the final part of architectural design activities.

Time can be crucial during home construction projects. For example, in cold weather climates, it is beneficial to start foundation and carpentry work before winter begins. If you are in a time crunch with winter closing in, consider using the Design/ Build System instead of the Traditional System, to beat the cold weather. Builders can find ingenious ways of compressing a construction time schedule when required. Another timesaving benefit is financial. Some homeowners are forced to carry the burden of paying two mortgages during construction, one on their present home and another on new construction in progress. Reducing the project elapse time can also minimize costly double payments. Time is money!

So do the advantages of the Design/Build System sound pretty good? Ready to go out and sign that builder up right now? Hold it! Just as in a legal trial, you need to hear both sides of the story before you make up your mind!

Disadvantages

• Disadvantages with the Design/Build System begin with a lack of competition. Since you are not soliciting multiple competitive architectural and construction bids, you will never know the real cost of your project. You are trusting your builder to quote a fair price. Without competition, why would the builder and his subcontractor sharpen their pencils? I have participated as an architect and a builder in both project delivery systems, and I know that without competition, prices will be higher. Since a home construction project is not a uniform product like a car, shopping builders at the beginning of the project is very difficult because so much of the project is still undefined. Could you sleep comfortably at night wondering if you are being ripped off?

• Another possible disadvantage of this system depends on when the price is determined. Builders are much like salesmen; they want to get your signature on a contract as soon as possible. On a custom project, this means you may be asked to agree in writing to a project cost before the design is completely determined. In their rush to complete a contract, builders are famous for neglecting to include all the items you might expect, or using inexpensive finishes below your expectations. For example, the contract calls for a cost allowance for light fixtures of $5,000. This might be fine for a middle-of-the-road project, but if you are expecting a ten-foot-diameter crystal chandelier hanging in your entry, your allowance will fall far short of covering the lighting fixtures for the rest of the house.

Any changes in design or upgrading of the quality of architectural finishes, such as floor coverings, appliances, and cabinetry will result in increased cost. You are again at the mercy of the builder to accurately reflect your level of design in his original price and to fairly charge you for changes. Regardless of how excessive his extra charges may be, if you want the change, you are obligated to pay. You have yielded a good amount of project control to the builder. Beware of signing a contract too early in the game until many of the design parameters are established.

• The third disadvantage in the Design/Build System is the lack of choice in selecting an architect. The builder, who supplies the architect, may offer only one or two candidates for your consideration. Keep in mind that the builder views the architect or designer as no different from a plumber or electrician. He is seeking an adequate job for the lowest price to maximize his bottom line for the project. If your project's design requirements are straightforward, and the Design/Build architect fulfills your expectations, then problems should be minimal. However, if you desire a high level of design, which you can't seem to communicate, or is not within the skill level of the designer, you are in for big problems. If you and the architect lack mutual understanding of the design ideas, you are likely to wind up with a project that fails to meet your expectations.

• The scope of architectural services is also generally less comprehensive in the Design/Build System than in the Traditional System. The builder will typically tell his architect, "Just give me enough drawings for a permit. I'll fill in the details later." These finer details are not shown on the plans so the builder must provide oral instruction to his subcontractors. This situation often causes communication problems between the builder and homeowner because the details are not in writing. Thus the architect or designer will only be present through the completion of the design. Their brief stay ends with the completion of the drawings. The builder considers the architect's attendance during construction unnecessary and will not pay additional architectural fees.

The architectural loyalties belong to the builder, not to you. While the design and drawing work proceeds, the architect is taking instructions from the builder, although you may think he is listening to you. The architect wishes to please his employer, the builder, for future work. He hopes to see his client, the builder, on many additional projects. You, he will probably never see again. Once more, you are trusting the builder to furnish a competent architect who will comply with your desires and will also specify quality materials. If you have an issue with the proposed design, the architect will follow directions from the builder, placing you at a disadvantage.

Even if the Design/Build architect was present during construction, he would be reluctant to report his employer's construction shortcomings to you. Therefore, you are alone in determining whether the builder is complying with the technical facets of the design, and is building within the industry standard for workmanship. The Design/Build System lacks the checks and balances of the Traditional System.

Bottom Line

Think about your project and then weigh the advantages and disadvantages of each delivery system. How much money do you have to spend? How much time can you devote to the project? How important are specific home features to you? If you have always wanted a home with a special design and are willing to spend considerable time and money to get it, then the Traditional System is probably right for you. If you will be happy with a project that has relatively standard features, or a limited scope, Design/Build may be the answer.

Remember that the Traditional System maximizes your control, but demands a great deal from you. If you use this system and neglect your project, you may lose control and end up with a problem. As I have explained, you must work on your new home from beginning to end. In Design/Build, you trust a builder with all the details of your project. You may never know the true cost of the project. Your design and finish options may be limited.

One final point. If the professional you choose has a terrific reputation, you will probably come out well in the end, regardless of the system used. Be sure to ask prospective project professionals about their project delivery system preferences. Some prefer to work within one system only. Weigh both their reputations and the delivery system they favor in making your decision.

Chapter Two Recap

- **Selecting the proper building team for your project is probably the most critical decision in producing a successful project.**

- **Understanding each participant's function and responsibilities in the building process is very useful. Architects, general contractors, builders and other supporting cast members play a very specific role in today's construction world.**

- **Two basic choices are available to the homeowner in selecting how professionals will provide their services.**

- ***Traditional System.* The homeowner hires two separate entities: an architect and a general contractor.**

- ***Design/Build System.* The homeowner hires one company, a builder, who is responsible for the entire project from design through construction.**

- **Each system has advantages and disadvantages.**

- **The Traditional System is good for advanced architectural design and increased competition on fees, but problems of architect/general contractor conflict and lack of cost control could surface.**

- **The Design/Build System offers cost control and time savings, but may lack competition and the checks and balances of an architect working directly for the homeowner.**

Chapter 3

Project Finances: How Building Professionals Set Fees and Services, and Funding Your Project

During a conversation on a recent flight the gentleman sitting next to me discovered I was in the building industry. Given this opportunity, he complained—in lengthy detail—about his construction disaster. Unfortunately, when you are sitting on an airplane, if someone wants to tell you a story, you are certainly going to hear it!

"I'm building an addition to my house, extending a new wing for an enlarged kitchen and formal dining room," he said. "The builder asked for 30 percent of the total cost as a down payment. He promised me that we would be enjoying Thanksgiving dinner in our new addition. I thought that 30 percent was quite a large deposit, but if he wanted that much, I figured he probably would finish faster. After the first few weeks he never had a crew working for more than two days running. Nobody came. All I got for my calls was a lot of excuses. When Thanksgiving came, we were lucky to have the openings in the roof closed before the first snow."

My flying companion made a mistake common to many homeowners. He did not understand how project finances should be administered. Because he was overpaid at the beginning of the project, the builder took his entire profit up front, leaving no incentive to complete the project on a timely basis.

Considering the high cost of construction projects, you need to become familiar with the different services each building professional offers and their fees. You will also need to know up front if you can afford the project you envision and when these funds should be paid. Our first topic will help you determine a preliminary price for your project.

Project Feasibility

Project feasibility is of paramount importance when you begin to explore pursuing a project. As you start visualizing the forms of the exterior and rooms, your biggest question is "What's this going to cost?" At this beginning stage without a definitive scope, determining construction cost can be difficult.

The most frequent question I am asked is "How much does it cost to build?" After years of going into lengthy explanations, I have now cut my response down to a question I ask in return. "How much does it cost to buy a car?" I love analogies with cars! "Do you want a new car or a used car? Would you like a Rolls-Royce, Cadillac, Chevrolet or a Volkswagen?" Like cars, home building projects range from the very basic to the luxurious. Costs will vary considerably according to your tastes, needs and budget.

It is actually far easier to price a car than it is to price custom construction. Other than available options, a car has a very definitive basic price, readily obtainable at a showroom down the street. Architectural design and construction is much broader, with no uniform sizes, shapes and pricing. Without reliable cost information, you could assume either too much or too little for project cost at this early stage.

First, our discussion is centered on building cost only. The price of vacant land or buying an existing house to adapt is far too variable for our general review. These are market-driven real estate costs, and should be readily accessible from other sources. Also, since architectural design costs are a fraction of construction, I will leave this factor for later.

Returning to our car example, the cost of the same identical model may vary 5 percent at most from all the dealers in the same area. I can take the exact same building project on the same piece of ground and receive as much as a 15 to 20 percent spread on construction proposals. If I take these same plans and compare construction pricing in New York City and Albuquerque, New Mexico, I could expect a whopping 25 to 30 percent difference. Each geographic region has its own cost of living level that impacts construction material and labor costs. A carpenter may be billed out at $70 per hour in New York, due to a scarcity of available union labor. The same carpenter in New Mexico is probably nonunion and charges $40 per hour. Already, we have a 40 percent difference in labor to construct the project.

Material costs can also vary. Contractors buy concrete from a ready-mix company. In New York, a yard of concrete sent to a site may cost $80 per yard because the driver has to sit in rush-hour traffic for an hour or two before he can deliver. Back out in Albuquerque, traffic is not so bad, so a half-hour trip to the site reduces the cost of that concrete to $60 per yard. There is another 25 percent difference. Get the picture?

Another regional difference is building code requirements. Certain big city areas may require copper plumbing piping, and electrical wiring run in conduit, a hollow rigid aluminum tube. Less urban areas permit plastic plumbing piping and shielded flexible electrical wiring. These two factors alone could account for a $20,000 difference in building a new house! So, if Uncle Max living in Florida tells you that new home in Boston should cost the same as his, beware! Unlike domestic cars mostly made in factories near Detroit, house projects are built everywhere under different economic situations.

Let's concentrate on local costs. Pricing a custom project is a subjective process. If I send a set of plans for an addition to three different carpentry subcontractors, I will get each sub's interpretation of how much material and labor he needs to complete the job. Their profit expectations will also be included in their numbers. Assuming all other factors, including quality, are the same, I am not surprised to see a 10 percent or larger spread in the subs' quotes.

With this brief introduction I have provided, I hope you see that construction pricing can be variable. When you first try to determine a preliminary project price, I encourage using numbers ranging from low to high. This will give you the best and worst case scenarios. The best bet for early cost calculations is using the *square foot method*. This represents the cost of every square foot in the total floor area of your project. New homes start at 1,500 square feet and head upwards to 3,000 square feet or much more, depending on your choices. Additions include only the new area being extended in determining the area. Remodelings count only the space to be altered. A word to the wise! The square foot method is most effective for new home construction, less accurate for additions and practically useless for remodelings. Figure 3-1 illustrates our cost estimating approach.

New homes contain a blend of expensive spaces, such as kitchens and bathrooms and cheaper space, like bedrooms, closets, hallways, and garages. The

Figure 3-1

Preliminary Construction Cost Evaluation Methods

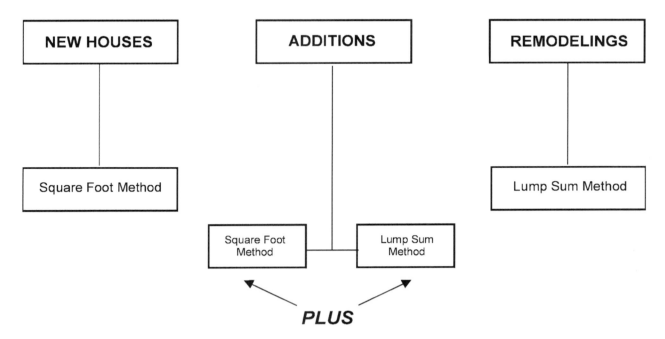

high cost of expensive spaces is offset by less expensive areas to produce an average. Most new houses will have an equal number of expensive and cheaper spaces. Thus, the square foot method is fairly reliable, if expressed in a range. I can take the same size house, same number of bedrooms and baths, built on the same site and drastically vary the cost by changing the building materials used. For example, I could specify brick for the exterior instead of vinyl siding. When choosing finishes, I could use wood flooring instead of carpeting, and wood windows in lieu of aluminum windows. These alternatives will greatly influence the square foot cost for the same sized project.

When beginning to talk with friends and building professionals about project cost, you must first determine the level of finishes as part of the square foot cost. The more expensive the materials, the higher the square foot cost. Regional differences across the country can also have a significant effect.

Returning to my frequently asked question about construction cost, my long explanation would usually cause the questioner to make a quick exit or follow up with "OK, but just give me an average cost!" Since I'm cornered with no escape, for new houses in the Chicago area, I will tell them $100 per square foot for a very basic design, $130 with several upgrades, $150 gets you a very nice consistent quality, and $175 provides something very special. Unless you can point out an existing house to a building professional and say "Copy that," a range of prices is the best you can get at this stage.

As I mentioned previously, estimating the cost of additions and remodelings is not best accomplished by the square foot method. For example, an addition consisting of a new large kitchen or a splashy master bath will skew the cost per square foot beyond comparison. The reason is that expensive construction spread over a small area produces a high square foot cost. Say a kitchen costs $20,000. In a

$300,000 new house with 3,000 square feet, that represents $6.66 per square foot of the total cost. The same kitchen in a $75,000, 750-square-foot addition results in $26.66 per square foot. With additions, I recommend a blend of per square foot cost and adding a lump sum for any kitchens or baths to remedy this unbalance.

Using our previous rule of thumb, I start with $100 per square foot for the basic 750-square-feet addition and add an additional $20,000 for the kitchen. This would amount to $75,000 + $20,000 = $95,000, or $126.66 per square foot. This blended method of cost estimating quickly brings some real numbers to your attention. It accounts for building the "shell," or building envelope, and compensates for the big hits of concentrated expensive construction.

Remodelings are a different animal altogether. Existing shell space is being re-used instead of adding new area. Therefore, the square foot area method no longer applies. Instead, use the lump sum method for estimating kitchens, baths or rooms for home entertaining with big-screen televisions and sound systems. You can assemble these costs by visiting dealers who specialize in cabinetry, plumbing fixtures, appliances, and audio/video equipment.

Before we leave this subject, one more word of caution. Other conditions can affect the cost of an addition or remodeling. Often a larger electrical service or a bigger incoming water supply pipe may be required to construct your addition or remodeling project. This is particularly true on older homes. These unexpected, nasty surprises can add big dollars to a project. Chapter 13 deals specifically with remodelings and additions, discussing the surprises you may encounter.

Services and Fees

The old saying "You get what you pay for" holds true for the most part in the building industry. Building professionals offer different services to produce projects, and they expect different fees. The basic difference boils down to the attention to detail and effort put forth. To demonstrate this principle, let's invent a mythical project. We will use the Traditional System as described in chapter 2 and hire two architects, Ace and Bingo, and two general contractors, Goody and Goofy, each working independently to produce the same project using two different paths.

Ace, our first architect, will carefully guide his client through the process, producing fifteen different drawings illustrating the entire scope of the project down to the color of the ceramic tile grout and the paint color of each room. He also builds a scale model of the exterior so the homeowners will understand how their project looks. Ace will further obtain contractor bids and visit the construction site once a week to observe the progress. I call Ace the "custom" architect because he's in the project for the long haul. *Bingo*, our second architect, produces the plans in record time but only shows his client some brief design sketches and furnishes five drawings. Bingo expects the homeowners and whoever builds the project to fill in a lot of the details not contained in the plans. His client lacks enough information to understand how the completed design will appear. I call Bingo the "production" architect.

Goody, the first general contractor, carefully examines the plans and asks the architect for several clarifications while preparing his bid. When construction begins, Goody has a full-time superintendent on site every day, coordinating and checking every aspect of the job. When the final payment has been applied for, Goody only asks for a few extras the homeowner added. During the warranty period, Goody has a roving tradesman that responds fairly well to repair requests. Let's call Goody the "comprehensive" contractor.

Goofy, our second general contractor, sends the plans to his subcontractors, takes their bids unchecked, slaps on his percentage markup and submits his proposal to the homeowner. When construction starts, he shows up at the site once or twice a week, allowing the subcontractors to build the project relatively unsupervised. When a subcontractor discovers that he forgot to include an item in his bid, Goofy tries to pass along the cost as an extra to his client, stating that if this item had been originally included, his bid would have correspondingly been that much higher. Goofy only returns for warranty work after receiving the first letter from the homeowner's attorney. I call Goofy the "knock it out" contractor.

Given this scenario, would you expect Ace the architect and Goody the contractor to charge the same as their counterparts, Bingo and Goofy? When our two projects are finished, you may have similar finished products standing side-by-side. Although Bingo and Goofy charged you less for the house, Ace and Goody kept you informed, answered your questions, and delivered quality construction inside and out. Price may not always be a good gauge for quality services, but expect to pay more for better attention.

As we discussed in chapter 2, you have a choice in procuring professional services. To assemble service and fee information for your project, speak with professionals for each delivery system. The following should be representative of what you will encounter.

Traditional System Architectural Services and Fees

I introduced our architects Ace and Bingo in our example. In the Traditional System, you will probably encounter both types, including licensed architects and unlicensed architectural designers. To simplify our discussion throughout the remainder of the book, I will refer to both as "architects." Custom architects' and production architects' services will vary along with their fees. I have participated as an architect in both classifications, working for private clients as a custom architect, and for residential developers as a production architect. Let me share with

you my experiences gained during twenty-five years of practice.

The first variable you will encounter is the scope of professional services each architect offers. As a reference, the American Institute of Architects (AIA) recognizes five distinct phases of architectural services. They encompass the scope of an entire project and are briefly summarized here:

The *Schematic Design Phase* includes conferences with you, after which the architect studies and analyzes the project requirements. He or she then prepares preliminary design studies, consisting of drawings, models, and other documents which illustrate the scale and relationship of project components. They are revised until an agreement on a general design is reached. When the homeowner approves the schematic design documents and a Statement of Probable Construction Cost is submitted by the architect, this phase of service is complete.

The *Design Development Phase* includes the preparation of more detailed design drawings and selection of materials relating to building appearance using catalogues and samples, heating, electrical systems and structural information. The schematic design is really perfected at this stage. The architect submits a refined and updated Statement of Probable Construction Cost.

The *Construction Documents Phase* covers the preparation of construction drawings and specifications describing in complete technical detail the construction contract work to be done: materials, equipment, workmanship, and finishes required for architectural, structural, heating, and electrical work. The architect also assists you in preparing information for general contractor bidders, proposed contract forms, and terms and conditions of the contract covering responsibilities during construction. The architect also makes adjustments to previous Statements of Probable Construction Cost. When the architect has prepared the construction drawings and specifications and has assembled the bidding documents, this phase is complete.

Only in the Traditional System does the homeowner, with assistance from the architect, get involved in the fourth phase, *Bidding and Negotiations*. This includes evaluating the qualifications of prospective contractors, obtaining bids or negotiated contracts, and reviewing the Contractor's Schedule of Values, a detailed cost breakdown of materials and labor.

Finally, only a Traditional System architect continues to work with the homeowner during the *Construction Phase* and *Administration of the Construction Contract Phase*. During construction, the architect provides the following services:

- Preparation of additional drawings to clarify the design.
- General administration of the construction contract includes periodic visits to the site to review the progress and quality of work and to determine if the project is proceeding in accordance with the construction documents.
- Review of the contractor's applications for payment, determination of amounts owned to the contractor, and issuance of certificates for payment.
- Preparation of owner approved change orders covering authorized changes in the project.
- Determination of the date of substantial completion, final completion, and issuance of the final certificate for payment.

Many of the terms above probably sound like the first day of a foreign language class. Don't worry. By the time you finish this book, you will be familiar with most. The custom architect usually prefers to follow a project path utilizing all of the above five services. He feels a project is best served with the architect controlling the entire process, from start to finish. You may feel that your job is too small or too ordinary to warrant this amount of attention. Regardless of size and complexity, every project can enjoy the benefits of full architectural services. Many

potential problems can be avoided by having an architect address all the design details, seek competition in construction bids, and represent your interests during construction.

Services provided by the production architect usually ends at the third phase: construction documents. He prefers not to provide assistance in bidding the project to general contractors, negotiating the contract, and observing construction on site. You are left to navigate these new waters on your own. I recommend that you at least check out the additional cost of these two services using a custom architect, and weigh the expense versus your confidence in self-administrating bidding and construction. Besides, during construction on every custom project I have encountered, questions invariably appear that only the project architect can answer. Having this continuous input by your architect can be invaluable.

There is also a distinct difference in the services provided by both types of architects during the first three phases. We mentioned earlier in our mythical project that Ace furnished an architectural model and considerably more drawings than Bingo. The custom architect takes additional time to help you fully understand how the proposed design will appear when completed. Three-dimensional aides such as models, computer-generated views, and hand-drawn perspective sketches make a world of difference if you are uncomfortable trying to understand flat, two-dimensional drawings. Because drawings in only two dimensions flatten or distort a three-dimensional image, the homeowner is frequently confused visualizing the final product. While these additional presentation tools are a great help, they are time intensive and demand greater effort and a higher fee from the architect.

The second disparity in architectural services, and the most important, centers on the design and specifications of material and interior finishes. While both architects accomplish many of the same tasks, much detail is left untouched by the production architect. Let's examine a list of these common and separate services. Some terms will be unfamiliar to you and will be explained in detail in chapter 10.

Services Both Kinds of Architects Provide

- Site plan of the house or addition on the lot.
- Floor plans with interior walls and doors.
- Basic indication of exterior and interior materials.
- Exterior walls and roof design.
- Structural system.
- Electrical power and telephone.
- Plumbing services.

Materials and Interior Finish Package Provided Only by a Custom Architect

- Exact selection of exterior wall and roof materials, including color.
- Flooring, including carpeting, wood or ceramic materials and colors.
- Cabinetry and counter tops.
- Plumbing fixtures and color selection.
- Lighting fixtures.
- Bathroom accessories, towel bars, etc.
- Special built-ins like fireplace fronts, closet storage systems, and cabinets.
- Appliances and equipment.
- Doors and door hardware.
- Decorative moldings, like baseboards, door and window trim, stairs and railings.

The above material and finish package can account for approximately 40 percent of a project's cost. The custom architect's scope of service usually includes selecting these items, while the production architect leaves the design and specification process up to you and the general contractor. This key difference will have a tremendous impact on your time and decision making. If you are working on a small addition or remodeling with a production architect, you may be up to the task, provided your general contractor is helpful. But on a new house project,

you can be buried with the amount of work that is your responsibility. Have you ever been to a developer's sales center where he has nice people showing you selections of carpet, cabinetry, and ceramic tile? Well, now you have taken on the job of assembling your own sales center for materials plus a lot more!

There is one more glitch caused by a lack of detail on your architectural drawings. As we have previously mentioned, when you give incomplete plans to general contractors for pricing, they have no idea what to include in their bid for these items. Don't expect them to be mind readers! This will cause major problems, as general contractors could plug in any amount, known as a cost allowance, for these unspecified materials. The result will be widely different construction bids. For example, one bidder included $7,500 for cabinets and another figured $12,000. This is one of the leading causes of disaster stories. You could be unsure about the cost of 40 percent of your project!

Now that you are acquainted with the differences in architectural services, you can guess that fees will also be all over the board. Architectural fees are on a pay as you go basis; the more you require, the more you pay. The custom architect and production architect usually set charges differently because their scope of professional work varies.

Fees for the custom architect using the full scope of services are usually based on a percentage of construction cost. For instance, if an architect earns a 10 percent fee on a $100,000 construction project, his fee would be $10,000. Percentage fees are preferred by the custom architect because his compensation equals the final amount of construction. This is a convenient method since custom project budgets tend to grow during design. If the $100,000 budget expands to $120,000, the architectural fee increases another $2,000. Renegotiating any additional fee with the architect is unnecessary because the adjustment is automatic.

You will find that architectural percentage fees can vary widely, from 4 to 15 percent of construction cost. Why would this be the case? First, fee levels are dependent upon the architect's prestige and reputation. The more well known and busy an architect is, the higher the percentage. Young architects starting out will lowball fees just to get something built. Older, more experienced architects aren't so desperate; they have a continuing supply of commissions. When I first started my own practice, my fees were very competitively priced so I could steal work away from experienced practitioners. I thought my first clients were getting a real deal. Now being older, I am often on the other end of the stick!

Secondly, the scope of architectural services varies. Not all production architects provide the same services, and the same is true of custom architects. Your goal is to try and assess the scope of services for each prospective candidate and compare them on an apples-to-apples basis. Once you understand exactly what each will provide, the difference in fees will be more understandable. Occasionally, some custom architects may bill their fees on an hourly basis, just as an attorney charges. Avoid this arrangement if possible; design is very subjective and without a limit, hourly fees can run wild. Always try to get a maximum limit on architectural fees. If your project scope is unclear with a variable budget, suggest to the architect that the percentage fee is acceptable, but regardless of the final project size, it will not exceed an agreed upon maximum amount. This amount should be less than the fee produced from their percentage times your maximum budget.

Production architects usually charge by the square foot area of the project, especially for new houses. Since they are often not involved with selection of interior finishes, construction cost is not viable as a basis for a percentage fee. Their basis is the project's size: the project floor area multiplied by a square foot fee produces a fixed amount. Again, you will encounter a range of square foot fees for many of the same reasons we mentioned in discussing the custom architect's fees. Geographic and experience factors also account for a difference. Architects with minimal background may charge $1 per square foot for a set of plans for a new house. More experienced practitioners with office staffs can charge from $2 to $3 per square foot. Smaller projects will command even more per square foot.

Another factor accounting for the variance in fees is professional liability insurance. Like doctors, architects can be found liable for their mistakes. Architectural malpractice insurance, sometimes known as *errors and omissions* (E&O) coverage, may not be carried by younger architects or smaller firms due to its high cost. E&O coverage is only available to licensed architects, so architectural designers are not eligible.

Few homeowners are aware of this insurance coverage. I can think of only a few instances in my residential practice where my clients have asked about my coverage. Yet this insurance can be very important if problems arise and the architect has limited resources. It is good to know there is an insurance carrier backing your risk. Since this coverage is expensive, insured architects usually will have higher fees to offset their expense. Gauging the risk of using an uninsured architect is difficult. Structural or weather-tight design errors can be made on any size project. The likelihood of mistakes may be less on a smaller project than a larger job, if this can be used as a guide. Even if architects are uninsured, they can still be held financially liable for their errors, but payment may be harder to collect.

Finally, architects generally require a deposit known as a retainer to begin a project. This up-front payment commits you to their firm and shows good faith on your part. The amount should be held only as a credit to be used for the final payment. Retainers usually do not exceed 10 percent of the anticipated total. If you do not finish the project with the architect, that portion of the unused retainer should be refunded to you.

General Contractor Services and Fees

A general contractor's project costs have many variables, from subcontractor materials and labor costs to business overhead, insurance, and profit. Assembling an accurate construction bid is a great deal of work for the many people that must be involved in the process. We have previously discussed the large number of subcontractors and their respective suppliers required for a building project.

Generals working within the Traditional System submit construction proposals based on the content of the drawings and specifications provided by your architect. Their bids are expressed as a *lump sum,* or total project construction cost. This number consists of *hard costs* and *soft costs.* Hard costs are direct costs to the general contractor for all materials and labor required to build the job. This would consist of all subcontractor bids plus any construction work provided by the general's own forces. Soft costs consist of business expenses: office and supervisory salaries, insurance, licenses, bonds, temporary electrical power, even tools.

First, let's discuss the hard costs. For most residential projects, a contractor may solicit only one or two subcontractor bids for each trade. The majority of his costs depend on the subcontractors' labor wages, the cost of building materials and their profit requirements. The more competitive the subs' bids, the more competitive the general's final price will be. Some subcontractors may have an ample supply of work and will not cut prices, causing higher general contractor bids. Other subs might be hungry for work and will quote more aggressive labor and profit rates resulting in lower general contractor prices. Basic economic supply and demand tells us that it's more expensive to build when the construction economy is hot, and cheaper during a recession. Unfortunately, few people have the money during a downturn to take advantage of these circumstances! As you can see, subcontractors' bid fluctuations can make or break a general contractor's success.

Soft costs are our second variable. Contractors operating a small business out of their home will have lower operational costs than their counterparts carrying the overhead expense of a bigger office and supporting staff. Is it logical to seek a general working out of his truck for the best cost? Remember, a contractor with no staff has no personnel for assistance. Again, you may well get what you pay for. A general's soft cost stays relatively stable and are easy for him to estimate. Once the subcontractor costs come in, a contractor typically adds a percentage of the hard cost to cover soft cost and profit. Usually the only way a contractor can reduce his price is to go back to the subs and ask them to reduce their profit margins, as well as reducing his own.

General contractors who receive a residential project often want deposits before the start of construction. They may cite the need to pay deposits on certain materials, like cabinetry, appliances, etc. This need is only partially valid. Most construction material suppliers have accounts that offer the general about thirty to sixty days for payment. The contractor is really looking for good faith, something to commit you to the project. If you are serious about starting construction, he wants to see some cash! A fair deposit is 10 percent; anything more is excessive. As my newfound friend on the airplane from the beginning of this chapter discovered, overpaying a deposit gives a contractor his profit for the entire project before the first day of construction. By the time his first set of material and labor bills are due, he will hit you for another payment to cover them. With no incentive for receiving profit as the job progresses, he has no reason other than his reputation to show up.

Design/Build Services and Fees

Fees and services for the Design/Build System described in chapter 2 are the most difficult to analyze because you often lack the requisite information. If you are seeking to buy a model house in a subdivision, you are given a total price. You have no idea of the land cost, architectural design fees, construction cost, or profit. The only analysis available for you is to comparative shop among other subdivision developments for perceived value.

You can face a similar situation using the Design/Build System for a new custom house, addition, or remodeling, because you will get only a total project cost from a builder. Some design/builders may be more willing than others to provide basic cost breakdowns for your analysis upon request, but without a detailed summary of subcontractor costs, you will only be able to compare bottom line numbers between builders. As mentioned before, a simple straightforward project should be fairly easy to compare. Larger, more complex projects containing many design decisions will be much more difficult. You therefore have two managerial tasks to qualify their bids. You must verify that all builder candidates are including the same design scope of work within their proposals, and you are obtaining a fair, competitive price. The only means of extracting this information is to ask a lot of questions. Once you have read chapter 8, you should be in a better position to do so.

A builder usually requires a nonrefundable deposit to begin preliminary architectural design and construction budgeting. This deposit is based on the project's size and can run into thousands of dollars. As a rule of thumb, it should not exceed 1 percent to 2 percent of the project and should be credited to the final cost. The design/builder in return should produce preliminary floor plans and a drawing of the front of the project, called an elevation, along with a written cost estimate. Since the drawings will be very sketchy at this stage, a supplemental written list of materials and interior finishes should also be provided. The more detailed this list is, the more comprehensive your estimate will be. Design/build fees for soft costs and profit should be similar to those charged by a general contractor. Prior to hiring a design/builder, you should complete your research and have confidence in the builder's abilities, as the initial deposit is usually nonrefundable.

Upon receiving these documents, you have reached a point of decision. In the event you wish to continue the project, another larger payment to cover the cost of completing architectural drawings, paying for building permits, and ordering long term delivery items is required. This payment should not exceed 10 percent to 15 percent of the project construction cost. If a land purchase is involved, this percentage will increase. If you are uncomfortable with the results of this first design phase, you should have the right to terminate the relationship without further obligation. Verify this provision before making a deposit.

Construction Finances

Building a new house requires financial resources to complete a large financial transaction. Even constructing an addition or an extensive remodeling project can be a significant investment. A basic understanding of construction financing and the cash flow process is a must for the educated homeowner.

Let's first discuss the amount of money you will need and how it is dispersed. If you are working with a builder who currently owns the property to be built on, he may finance the construction, requiring you only to secure a conventional mortgage when you move in. He is providing the construction loan and you are paying the interest in the final price. If you own the property and are using either the Traditional or Design/Build systems, you will be required to have the funds to pay the professionals during design and construction. Determining your financial resources and limitations is an important step you should take early in the process.

Although there are many ways for you to accomplish Traditional or Design/Build system project financing, the most common method involves two loans. A *construction loan*, a short-term loan usually tied to a rate above the current prime interest rate, is established to fund the construction. As construction progresses and the contractor/builder submits requests for payment, you draw upon this loan. The longer construction lasts, the more monthly interest payments you will make. Upon completion of construction, a permanent conventional mortgage, also known as an end-loan or take-out loan, pays off the construction loan and becomes your long-term mortgage, payable every month. Since the construction loan is not amortized over twenty to thirty years, the rates for monthly payments are higher. This is a tempting incentive to finish building quickly; it is much more economical to pay a mortgage payment than a short-term construction loan payment.

The architect, contractor, or builder will submit bills for payment, usually on a monthly basis or as certain *project milestones* are reached. Milestones could be the completion of the design phase by the architect or finishing the house framing by the contractor/builder. Either you write a check from your own personal resources or apply to the lender to fund payment through your loans.

In today's ever expanding mortgage market, there are several sources for securing the financing you need. A conventional bank will give you the short-term construction loan, but may not be interested in the permanent mortgage financing. Other lenders may only be interested in the permanent mortgage and not the short-term construction loan. You may be required to find two different financial institutions to fund your project.

To grant a construction loan, the lender evaluates specifics common to all loans: your ability to pay the monthly charges, the value of your collateral, (namely your current home and your new project), and current assets. Arriving at a value for a new home to be constructed requires an appraisal conducted by the lender. An appraiser who is familiar with real estate values in your area will need at least a preliminary set of architectural plans to assess the size and quality of the new construction. Often the lender will also require a written proposal for construction from a contractor/builder to evaluate with the appraisal. This appraisal value and your credit worthiness will determine how much the lender will finance.

Typically, construction loans will fund only 70 percent to 80 percent of the appraisal value of the project. If the construction cost exceeds the appraisal, which often happens with additions and remodelings, the lender will always use the appraisal value for the loan, causing an additional shortfall for you. To make matters worse, you are required to provide your portion of the money first, before the lender will begin funding the loan! You must show proof that you have first paid your portion to the building professionals.

Many homeowners have their funds tied up in the equity of their present home, which is not available until that property is sold or refinanced. Since they do not wish to sell and close on their present home until the new project is complete, where will they obtain the 20 percent to 30 percent required? Another loan, known as a *bridge loan*, can solve this problem. A *bridge loan* is underwritten using the equity in your present home as collateral. It is another short-term loan usually made by a conventional bank to make funds available for the homeowner's required 20 percent to 30 percent. A home equity loan could also provide the same result. Again, you pay monthly interest payments. Once construction is complete and the project is occupied, the permanent mortgage will pay off the short-term loans previously secured for down payment and construction.

Your best bet to economize on project financing is find one lender to provide both short and long term loans. By hunting all over town or using the Internet for a broader choice, you can find a bank or other lender that will offer a complete package deal. Mortgage brokers are another source worth considering. This enables you to save on duplicate loan fees and other service fees for application, appraisal, and credit checks. This can amount to thousands of dollars on big projects.

Another source for potential savings involves the title insurance company used for construction payments. A title company is commonly used by the lender to administrate payments to the contractor. Each time a payment is processed, the title company charges a fee. Try to limit your contractor/builder's number of construction payments to limit your title company fees. Three payments for the entire project is an ideal number for this concept. This tip alone can save hundreds or thousands of dollars, money you can use for better things!

As discussed earlier, you want to control the quality and schedule of your project. Dispensing project funds is a terrific leverage tool to accomplish this goal. When and how much to pay is a topic we will explore in depth in chapter 11. Project finance is your companion throughout the project, and one that we will be frequently revisiting.

Chapter Three Recap

- **Determining if a project is feasible starts with establishing the approximate cost of construction. Even though many specific details of the project are not developed at this point, several methods are available for "ball parking" a cost range.**

- **The *Square Foot* method is useful for new houses and assumes a unit cost per square foot multiplied by the house area. Since construction costs vary widely across the country, the homeowner must use specific local prices.**

- **Additions are best approximated using the square foot method for the new space, plus adding sums for expensive individual spaces, such as kitchens and baths.**

- **Remodelings are so variable in scope that only the sum of even individual trade should be used as a cost basis.**

- **Architects, general contractors and builders do not provide the same scope of services. Architects will vary in the amount of design work provided and general contractors and builders administrate the construction site with varying levels of supervision.**

- **When comparing fees and costs among competing building professionals, the homeowner must carefully analyze the professional services provided and the corresponding cost.**

- **Residential construction finance should be considered like an investment. Economizing on lending costs can provide valuable dollars to be spent elsewhere.**

Chapter 4

Acting as Your General Contractor—How Hard Can It Be?

Years ago I designed a new house for a couple who thoroughly enjoyed the design process, although they had an ambitious architectural program with a limited budget. As it became apparent through estimates and actual contractor bidding that their budget was insufficient to pay for their desired design, my client suggested he act as his own general contractor. "If I can save a general's markup, we can afford to build the project without having to cut back our wish list," he reasoned. I agreed that the money saved would make the difference, but cautioned him that he was a dentist, not a contractor, and didn't know the first thing about construction.

"How hard could it be," he replied. "After all, I'm a successful professional. I run my own business and certainly have the time. Besides, if you help me, I know I can make it work. You know the subcontractors and the nuts and bolts. Together we can't miss!"

In spite of my warnings, he decided to go for broke. I could now foresee my typical day for the next nine months: two to three phone calls per day, explanations of why a plumber doesn't install telephone wiring, and messages from my subcontractor friends asking me if I was out of my mind. To make a very long story a lot shorter, my client invited me over for dinner after his family moved in. "Never again," he said, "I couldn't believe how much work this

turned out to be! Some of my subcontractors lied to me and never showed up on time. I probably ended up spending as much if I had used a general contractor, and my dental practice really suffered."

Homeowners occasionally want to act as their own general contractor. A trip to the local library will arm you with dozens of books on the subject. The books proclaim that paying a general contractor all that overhead and profit would be money better off in your pocket. Just like my client, most self-contractors seem to be successful people who think that home construction is only one more challenge they can conquer. Many romanticize the idea as an adventure, like scaling Mt. Everest without any prior climbing experience. Well, I am here to try to change your mind, if you are so inclined. Our discussion can accomplish two things: it will briefly acquaint you with a general contractor's job and also try to talk you out of changing occupations!

Let's review what a general does for his money. I am going to skip a lot of the obvious chores and just highlight some of the lesser-known issues facing a contractor in today's world. To simplify my writing, I will use the pronoun "he" when referring to a general contractor. This convention is no longer correct, because women are appearing in all types of construction occupations these days.

Subcontractors

By examining the project plans, the general determines what materials are required and the associated subcontractors who will be needed for their installation. He must be familiar with all trades' scope of work, what they can and cannot do. Beyond the basics that carpenters pound nails and roofers install shingles, there is much more to know. Today's construction world is very specialized. For example, water softening systems are not installed by plumbers; installers of water treatment are hired separately as a specialty trade. Telephone and home entertainment wiring is not a part of the electricians' job; they sub that slice of the project to another specialist. Concrete subcontractors don't touch exposed stone patterned concrete, that is someone else's work. At the end of a high-end new house project, more than twenty to twenty-five separate subcontractors will be involved in completing the project.

The average homeowner has no knowledge or training regarding where one trade finishes and the next begins. With all the specialties in today's market, the general contractor must know how to evaluate each subcontractor's written proposal for thoroughness. Each subcontractor should include every specific task in a comprehensive written contract to avoid confusion. Unfortunately, subs often abbreviate their proposals to be very generic. For instance, an electrician may simply specify "electrical work in accordance with the plans." Sounds like everything would be included, right? Maybe not. When he shows up on the job the first day to review the installation, the general could discover the electrical contractor did not include furnishing and installing the outdoor patio lights or wiring the hot tub on the exterior deck. Another example of incomplete subcontractor proposals is buying a cabinetry package only to discover that installation was not included.

This process of analyzing subcontractor proposals is called qualifying bids. Through previous experience, the general knows what questions to ask when a sub's vague proposal is submitted. The concept is to verify that every item required to complete the project is covered under someone's scope of work. If the general overlooks an item that he assumed a sub included, there is a good chance the extra cost will come out of his own pocket. The more mistakes the general makes, the less profit he realizes. This is tough duty for a novice homeowner to accomplish. Even the pros frequently make these errors. If you are anticipating that the original total cost you used for the construction loan would cover all expenses, you are now in for a nasty surprise. If your loan is too low to cover all the costs, be ready to scrounge for more money.

Another facet of qualifying subcontractor bids is determining if the proposed price is fair. It would not be unusual for subcontractors, knowing that they are quoting a job for a rookie, to inflate the price. They may even see a risk of doing business with an inexperienced homeowner with no track record. How can they be sure they will get paid? The homeowner-contractor does not have to worry about his reputation to get the next job! My guess is most subcontractors will boost their price by 10 percent to cover both of these circumstances. This additional premium already covers at least half of a general contractor's fee.

One of the hardest parts of self-contracting a project is to locate the right subcontractors to build the project. Homeowners usually assume that these tradesmen are located in the telephone yellow pages. This would be an expensive assumption, because the yellow pages contain service tradesmen, whose chief occupation is repairing existing construction, not building new systems. Construction subcontractors typically are not listed because they work for general contractors, not the public.

There is a wide difference between these two entities. Service contractors usually charge a high rate by the quarter hour plus travel time. Their jobs are geared to last only a few hours and several stops are planned for the day. Everyone has had the experience of calling a plumber for an overflowing toilet or clogged drain. You hold your breath until the bill is presented. Construction subs on the other hand quote a fee for a total installation at a much lower hourly rate. If a general contractor used electricians, heating, and plumbing companies that are all service type organizations to build his projects, he might charge 100 percent more for construction.

I have known many construction subs who moan that they should become service companies, with higher pay and fewer headaches! Many of the larger subcontractors actually have two separate divisions, service and construction.

The usual method of finding construction subs is by word-of-mouth. Generals will often pass along the names of good subs as well as bad ones to fellow contractors. Without having access to this network, the self-contractor is in for a hard time locating good construction subcontractors who will work for a fair price. This issue alone can cancel any cost savings advantage over using a general contractor.

Regulations

We should also discuss the legal requirements facing a general contractor. As I mentioned in chapter 2, many states and local municipalities require generals to be licensed. Often this license is only obtained by passing a written exam. Extensive knowledge of the building process, technical issues, building codes, and construction law are at the core of a challenging test. The inexperienced homeowner has virtually no chance to pass this Olympic test. Some governmental units used to waive the license requirement if a homeowner was constructing his own project; however with liability now being a chief concern of building departments, this exemption is rare. This issue alone can stop you before you leave the starting blocks.

I remember hearing a third-hand story of a self-contracting homeowner who applied for a build-

ing permit without a general's license. He managed to convince a sympathetic building permit clerk to overlook the license requirement. About one third of the way into construction, a building inspector from the department came to the site to conduct a regularly scheduled structural framing inspection. The eager to please homeowner struck up a conversation with the inspector, instantly displaying his ignorance. When the inspector casually discovered that the self-contractor had no license, he immediately shut down the entire job from further construction.

"You can't build without a license in this state! I don't care if you electrocute yourself when you move in, but if you ever sell this house, someone else is going to have to live with your mistakes. Either get yourself a license or find a general contractor who has one." Now this poor guy was in a dilemma; he had no chance of passing the exam, and no general he spoke with was interested in getting involved this far down the line. "How can I guarantee the project when the work was already done by someone else?" one general stated. "No thanks!"

Another component of government regulation is the volumes of building code requirements a general contractor must know. Today's codes regulate every building material and installation procedure. For example, the code specifically tells a carpenter the size and number of nails used to fasten a wood wall stud to a floor plate. Electricians are instructed to use exact wire sizes to feed different electrical loads. The code limits the number of connections a plumber can attach to a pipe. These regulations run for hundreds of pages.

You may be thinking at this point, "So what! The plans should include all the required information." Unfortunately, this is not the case. As I am preparing architectural plans for any project, I don't have the time to spell out every code requirement. If I tried, the plans would run hundreds of pages, which no one would read. Most architects will only highlight the important regulations usually required on the drawings to get a building permit. The remainder is handled with one note on the front page that usually runs like this: "The general contractor shall conform to all prevailing building codes in constructing the project." This simple sentence takes the responsibility from the architect and places it directly on the contractor's shoulders.

I bet your next thought is, "Well, if many of the subcontractors are licensed, they must certainly know the codes." This is a good thought; unless they are right off the boat, most subs do know the codes. But problems can still occur, for instance when the carpenter is nailing up the wall framing on Friday and notices that the nail supply is running low. Instead of running to the hardware store for more, he gets the bright idea to install only two nails in each stud instead of the required three. "I'll stop Monday morning for more nails and add the third nail later," he reasons. However, next week comes and either he may unintentionally forget or he may conveniently overlook last Friday's subpar work. A sharp general contractor will be able to keep tabs on his subcontractors, making sure all work is built in accordance with code. Unless a self-contracting homeowner is well versed with the regulations, he stands little chance of knowing what to look for.

The homeowner who decides to contract his own house construction is going to develop new friendships at the building department. Obtaining a permit these days involves more than just submitting the drawings. An application listing very specific, technical information is usually required, stating everything from the volume of the new project to the property tax number for the parcel. You must plug in the appropriate information when asked the electrical service size, the amount of air-conditioning capacity, the foundation system, and framing classification. The general contractor knows all the answers, so this is a fairly quick application for him to complete. But lots of phone calls and trips to city hall await the self-contractor to compile all this informa-

tion. Most of the important subcontractors must be listed, including their license and bond numbers.

The permit review process does not end with the application. After checking the drawings for adherence to the codes, it is common for the building department to require corrections or even changes to the design plans. This requires getting the architect re-involved, with numerous trips back and forth between the architect's office and the building department. For example, the code review comes back stating that the bedroom windows are too small to provide the required natural daylight and ventilation. The plans must be revised, adding larger windows. No big deal? Well, your window package price just increased, not to mention that the window order you already placed is incorrect. While this is just another day at the office for the general contractor, the homeowner will have to get time off from work to take care of all the details.

Taking the preceding a step further, many noncode project requirements are not shown on the architectural drawings. Typically, my drawings don't instruct the general to arrange for temporary power connections, remove extra dirt from the excavation off-site, or conduct his daily construction work in a safe manner. There are probably hundreds of procedures required by a contractor to keep a project running smoothly on a daily basis. Suppose the self-contractor hits underground water when digging the foundation? The water fills up the basement partially dug by the excavator. "Well, I can't go any further until you pump this out," he says. When the homeowner-contractor asks if he has a pump, he replies, "Nope, the general always does that for me."

Now the self-contractor is faced with renting a pump and large diameter hose to evacuate the water. Guess what? The pump runs on electricity and the power company is not due for another three weeks to drop in a temporary electrical service. So a portable gas powered generator is added to the shopping list at the rental store. As the pump finishes clearing out the water, the self-contractor notices that the darned excavation is filling up slowly with water again! He will be pumping until the concrete foundation has been completed.

The experienced general contractor has an item called "de-watering" when compiling his bid. He knows there is always a chance for underground water to appear and he wants his bid to include the money to cover the pumping expense. This is just one example of "contingency conditions" used by a general contractor. He has additional funds reserved within his bid to cover unexpected expenses. The self-contractor paid for the two-week equipment rental out of his pocket because he did not foresee adding a contingency amount to his construction loan.

Financial and Insurance Issues

Since I have mentioned construction loans several times, a good topic for discussion is funding the project. Financing your project as your own general contractor can be difficult. Many institutions do not feel comfortable risking their capital investment with an inexperienced novice contractor. Their concerns are usually twofold. First, the homeowner-contractor is typically unqualified to adequately fulfill the technical obligations of a general. Why would a bank want to underwrite a loan when there is a chance the construction work will be substandard? A finished building project is usually the collateral that guarantees the amount of the construction loan. If the loan defaults for any reason, the bank may have an unsafe house that's monetary value is less than required to cover the loan amount.

Secondly, financial institutions view the self-contractor as a one-man band. Let me share my own story in this regard. Years ago I bought an old hay barn in the country to convert into a house for my family. The location and property were nearly perfect, but the barn, built around 1915, needed a lot of work. Blissfully happy to locate such a find, I signed the real estate contract, figuring that a construction loan should be no problem. After all, I was

an architect, although rather young. Every local bank turned me down, some saying renovating that old ramshackle barn was very unconventional. I almost had one small banker convinced, though. He liked the idea and the numbers seemed to work, but one issue bothered his loan committee. "If you break a leg halfway through construction, or even worse, a car nails you going around a corner, who will finish the project?" he asked. "You have no one else to fall back on to complete the job. Sorry, but we can't take that risk. If you hire a general contractor, come back and see me."

By this time, I was past the point of no return on the real estate deal. I had no choice but to proceed. Using the funds I got from selling my existing house, plus personally borrowed money, I rebuilt the barn to near completion, far enough along to finally secure a loan. I found out the hard way that financial institutions like conservative, sure bets when they hand out their money!

While, we're discussing paper-pushing topics, let's talk about contracts, payments, and insurance. General contractors spend as much time in the office handling the business end of projects as they do supervising construction at the building site. You would be amazed at the amount of paperwork a project generates. First, each subcontractor working for the general requires some type of written contract. The easy way for the self-contractor is to sign a sub's one-page proposal which could be short on specifics. But as you will discover in chapter 5, construction contracts require a great amount of detail to cover all the bases. A sharp general contractor will have a comprehensive contract with all his subs, covering issues such as insurance, payment, defects in the work, and default of the contract, just to name a few. Since the self-contractor may not know what should be in the contract, he often fails to include these essential agreements in his contracts with the subs. Without a comprehensive contract—for example, one which requires insurance—even a simple injury at the site can become very expensive when a personal injury attorney comes calling.

Today, insurance is an important aspect of contracting. Generals carry several types of specialized insurance, covering general liability, injury, vehicle, and Worker's Compensation. Protection afforded by insurance is very important. As Murphy's law tells us, if something can go wrong, it will! Insurance companies are a lot like banks. They also like conservative, safe risks when extending coverage. An established general contracting company is a much safer bet than a rookie self-contracting homeowner. Many carriers will refuse even to consider insuring this type of risk; there is no prospect for continued future business and the odds are much higher that a claim will be filed. If the homeowner-contractor can get the insurance, expect the premium to be costly.

The largest amount of time a general spends in the office relates to getting paid. As you'll learn in chapter 11, the construction payout process requires paperwork involving all the subcontractors and their material suppliers. Since real estate title companies are commonly involved distributing payments, every number on all these forms must add up to the penny. No room for sloppiness here! Many general contractors who build several projects simultaneously have someone sitting in the office taking care of the paperwork, full time. Considering that many material suppliers want to be paid monthly, new payment requests often follows the preceding applications without a break.

A seasoned general has the procedure down to a science; his subcontractors and suppliers know exactly what paperwork is expected. The first-time self-contractor is in for a steep learning curve. Expediting paperwork is not the strong suit for many subcontractors. Often it takes several attempts just to get the right information shown on the correct form. I guarantee that the homeowner-contractor will soon be on a first name basis with the staff at the title company. He will be making frequent return trips

because the subs' paperwork is incorrect. Since everyone likes to get paid, the self-contractor better figure it out quickly.

Managing Construction

Up to this point, our discussion focused mostly on activities prior to the start of building. Day-to-day construction requires a tremendous amount of planning and coordination by the general contractor. Building a residential project follows an orderly schedule of tradesmen and materials arriving at the site at the proper time. Since construction is a cumulative process, starting with the foundation and building upward, one trade's work is dependent upon the completion of work by the previous trades. The carpentry sub can't begin to frame the house without a foundation in place. The roofer will not install shingles without the roof structure first being completed. These are fairly obvious, but how about these questions. Who installs first, the plumber, electrician or heating subcontractor? Does the wood flooring or cabinets come first in the kitchen? Is the insulation supposed to be in place before or after the building department rough-in inspection?

The average homeowner has no idea how to answer these questions. Actually, there are no correct answers to these three questions; answers depend on the individual project. If construction is expected to be completed in an expeditious manner, the subcontractors and their materials must adhere to a comprehensive schedule. Installing work out of order will guarantee time delays and problems down the road. Even the most experienced general contractor has difficulty keeping the process on track.

Subcontractors can also be notorious for failing to appear at the proper time. Either they spent more time than expected on the previous job, or the rain delayed their work and they are behind schedule. Equipment breakdowns and injury are also common occurrences. In some states, construction actually comes to a halt at the start of hunting or fishing season! There are dozens of reasons why subs don't show, but the consequence is delay for every trade scheduled to work downstream. Each newly arriving trade depends upon completion of the previous subcontractors' work. Subs are not going to wait around until absentee workers finally show up. They are going to the next job in line saying, "Call me only when you're ready." This is the most frequently heard phrase on a job site. Generals call this activity "chasing subs."

So, despite all these problems, let's say the self-contractor has made it to this point still in his right mind. Work is moving along, when the drywaller complains that the carpentry framing is so out of square, he can't possibly install the drywall. "The corners are so bad, I'd have to taper every sheet to make it work. Call me when you're ready with square corners." As he leaves in a cloud of dust, the self-contractor is left with an empty job site, wondering what to do next.

Most homeowners are not equipped to judge the quality of construction work. Many trades depend upon the workmanship of previous subcontractors for their own installation. For example, if the foundation is installed out-of-square, the carpenter will have a difficult time making his framing work. If the drywaller does a substandard job on the taping and sanding, the painter may not accept the surface to apply his paint. There are dozens of cases where a good general contractor will critically observe construction to avoid these overlapping problems.

As we have mentioned before, an ongoing relationship exists between a general and his subs. If one trade does a poor installation, or fails to correct faulty workmanship, he is risking future employment. He is much more likely to fix his problems for a general than for a onetime self-contractor. In our example of the drywaller walking off the job, it is the homeowner-contractor's responsibility to confront the carpenter, point out the problems, and ask for repairs. The construction business is not known for tradesmen who are cream puffs. When questioned about

the framing, the carpenter would likely say the drywaller doesn't know what he is talking about. He'll insist his carpentry work is right on the money.

Part of a general contractor's job is confrontational. Some subcontractors are famous for whining about almost everything. So the general must decide on one of the following approaches: To the carpentry sub, he could say, "Fix the blankety-blank corners or I will bring in another carpenter who knows what he is doing, and I will deduct the cost from your contract." Or he could tell the drywaller, "Stop your complaining. The framing is fine the way it is, just make it work."

If you are interested in self-contracting, do you like confrontation? Can you be firm and make good on a threat to withhold money? Remember, if you are a first-time self-contractor, your subcontractors will all be strangers and could tell you to take a hike. If the dispute cannot be settled, the finished product will certainly suffer.

I have saved the best for last. The largest demand will be on your time. General contracting is a job requiring full time and overtime. A general is always on the construction site at least once a day, and often two or three times per day as the project is finishing. If you're a self-contractor who has an inflexible full-time job, you probably won't have enough time to spend on the project. In addition to visiting the site daily, you can expect to spend plenty of time on the telephone. Unless a certain sub is working at the site, the only way to reach him is in the evening. It is not unusual for a general contractor to spend a portion of every evening on the phone, trying to coordinate their subs' activities. Since most do not have offices, but operate out of their homes, communication can be tough. I can still remember from my contracting days the number of times my call messages were mislaid by a sub's child answering the phone!

Now that I have presented a few of the daunting tasks performed by a general contractor, how much does he charge for his services? As we discussed in chapter 3, a contractor will normally receive 10 percent to 20 percent of the direct project cost to cover his overhead and profit. When a homeowner looks at this cost on the bottom line, it jumps off the page as a big number. When homeowners commit themselves to a large investment, the prospective self-contractor must weigh the cost savings advantage versus the disadvantages we have discussed, not to mention the resulting stress of accepting this challenge.

I have found that residential projects take on a life of their own. They will become a part of your existence for a while and soon take over. Projects also seem to develop a momentum. Those that have an effective design phase seem to go well during construction. Projects that start off poorly seem to stay that way to the very end. It is much like raising a child. For a short time either you will enjoy the process or wonder why you ever bothered!

In the next chapter many of the new concepts will apply to our discussion on self-contracting. We'll discuss why simply signing contracts to get your project started never guarantees a successful project. You must spend a good deal more time on contracts than it takes to sign your name. You'll also learn the specifics of several kinds of insurance you'll need before you embark on your building project.

Chapter Four Recap

- **Homeowners should familiarize themselves with the responsibilities of a general contractor before they choose to construct their own projects.**
- **General contractors must be familiar with many facets of construction.**
  - **How each trade works**
  - **Finding the best subcontractors**
  - **Coordination and scheduling of materials and labor**
  - **Knowledge of building codes and construction practices**
  - **Contracts, insurance and construction safety**
  - **Financial transactions for payment**
  - **On-site supervision**

Chapter 5

Why You Need to Check Contracts and Buy Insurance

As most homeowners are aware, a new house project, addition or remodeling may be the largest financial transaction of their lifetime. With a great deal of money at stake, homeowners need the protection of a comprehensive contract or agreement to protect their interests. Contracts are usually viewed by the public as a necessary evil, much like insurance policies. No one reads the fine print unless they need to file a claim. Then they pore over every word, wondering why they didn't check the language more carefully when they bought the policy. If a problem arises on your project, you don't want to find yourself grabbing the contract you never fully understood to review the fine print.

I read a newspaper story a few years ago that illustrates this point. A couple who saved for years to build their dream house were finally in a position to take their first step. They admired a house in another section of town, which had the style, layout and amenities they had always wanted.

Tracking down the original builder, they asked him to duplicate the house on their lot with a few minor changes. The builder quoted a price; they made a few design adjustments and concluded with a handshake agreement. The couple was so thrilled, they threw a party on the vacant lot! Just before ground

breaking, the builder sent his contract in the mail for their signatures. The agreement ran well over ten pages, but the price listed matched their previous estimate and the builder certainly knew the house they wanted, so they eagerly signed, ignoring most of the legal jargon.

Halfway through construction, the couple noticed that work had slowed to the point that hardly any tradesmen appeared at the site. Their calls to the builder's office were never returned. Finally, a friend told them she had heard the builder had defaulted on a large development project across town and had filed for bankruptcy. Racing to an attorney's office with contract in hand, they discovered the agreement they cheerfully signed actually restricted their legal actions and remedies against the builder. They could neither force him to complete the project or return their deposit. The attorney told them they faced a costly, lengthy, legal battle with an uncertain outcome in order to complete their dream house.

The lesson from this unfortunate story should be clear. Although you will have to spend some money on legal fees, contracts should not be signed without consulting an attorney. This cost is cheap insurance compared to the expense of litigation, or worse, having no rights at all. Without an effective agreement, you are at the mercy of prevailing state law or the whim of a judge or jury.

Proposals Come First

The first step in launching a building project is understanding the agreement process a project usually follows. Rarely do building professionals initially offer a contract. Contracts are a lot of work. Instead, they will first provide a *proposal*, an abbreviated document stating basic terms and conditions for their services. Building professionals want to see whether you are serious about using their firm before going through all that paperwork. Since proposals are commonly used, you need to understand the strengths and weaknesses of proposals used in the building industry.

Proposals and contracts between a homeowner and architects, contractors or builders are two different documents and should be used to establish working relationships at different phases of a project. As you qualify and interview potential candidates for your project, insist that the services and fees quoted by each firm should be formalized in a written proposal. Once you have selected the firm you will work with, the proposal should be converted into a separate, comprehensive written contract.

Proposals are useful written agreements prepared by building professionals at the time you are comparing services and fees for final selection. Figure 5-1 illustrates a proposal from an architect for project services. They can come in all sizes and shapes, from a preprinted single page form, a letter, or even a note written on the back of a business card. Since proposals come in many styles, you have the challenge of comparing competing proposals for specific services and fees. Proposals that are too brief and undefined will require you to request the preparer to provide additional information. If you are unsure exactly what a building professional is proposing and no additional explanation is offered, consider eliminating them from further consideration.

Don't make the mistake of relying on someone's good reputation or a great recommendation, then skim quickly over a proposal. Every aspect of the project, including services and fees, should be stated in writing within this document. Regardless of their format, they should contain the following information:

• Identification of all parties involved in the primary project agreement.

Besides your name, you want to know the actual legal entity of the company submitting the proposal. Although many companies regularly use their legal name, some may have a hidden corporation or partnership that is the truly responsible company. This research may divulge yet another company name whose history you will have to explore. Occasionally, a company will hide its questionable past by using another name until contract time, only to spring a surprise after you are far down the project road.

• Brief description of the scope of work for the project.

The specific nature of the project should be listed. For example, "a new two-story, 3,000-square-foot house, with two-car garage, four bedrooms, two and one-half baths. . . ." or "an addition consisting of 800 square feet for a master suite with bathroom and the complete remodeling of the existing kitchen." Everyone should clearly understand how much work is to be completed.

• Specific services to be performed

Every service to be provided should be itemized in the proposal. For architects, services are usually related to the five distinct phases of architectural practice we previously discussed in chapter 3. The proposal should state which of these services are included and which are excluded. Specifics are very important. For instance, if an architect proposes to visit the site during construction, the actual number of visits should be specified. The clause could read "weekly visits for the length of construction" or "a total of ten site visits."

Richard Preves & Associates, P.C.

Architecture   Planning

Figure 5-1

Architectural Proposal

February 29, 2000

Joe and Mary Client
1234 Main Street
Somewhere, US 12345

Re: Proposed Residence

Dear Joe and Mary,

PROJECT UNDERSTANDING:

We understand that you intend to build a 3,500 square foot residence on your five acre lot in Somewhere, US.

BASIC SCOPE OF SERVICES:

We propose to provide architectural services for this project described in the following Basic Scope of Service:

A.   **The Schematic Design Phase** includes conferences with the Owner after which the Architect studies and analyses the project requirements. From these we prepare schematic design studies consisting of drawings and other documents illustrating the scale and relationship of project components, including such considerations of structure and materials as may be appropriate at this time. Three-dimensional modeling shall be prepared for the Owner's review. Upon approval by the Owner of the Schematic Design documents and a Statement of Probable Construction Cost and Project Schedule itemized trade-by-trade submitted by the Architect, this phase of service is complete.

B.   **The Design Development Phase** includes the preparation of more detailed drawings and other data relating to building appearance and structure, mechanical and electrical systems, construction materials and finishes. Finishes, including all permanently installed items, such as plumbing and electrical fixtures, appliances, cabinetry, floor coverings, ceramic tile, and hardware, shall be viewed and selected by the Owner with direction from the Architect. Non-permanent finishes such as furniture, wall coverings and window coverings are not included. Coordination of tasks and schedules of the Owner's Landscape Designer shall be addressed. The Architect also submits a further Statement of Probable Construction Cost and Project Schedule. When the Owner approves these documents, this phase is complete.

February 29, 2000
Joe and Mary Client
Somewhere, US 12345
Page Two

C. **The Construction Documents Phase** includes the preparation of working drawings and specifications describing in technical detail the construction contract work to be done; materials, equipment, workmanship, and finishes selected by the Owner required for architectural, structural, plumbing and electrical work and related site work. The drawings and specifications, to the best of the Architect's knowledge, shall meet the local building codes and industry standards. We also advise the Owner of any adjustments to previous Statements of Probable Construction Cost. When the Architect has prepared the working drawings and specifications, this phase is complete.

D. **The Bidding or Negotiation Phase** includes advising the Owner about the qualifications of prospective contractors, and assisting in obtaining bids or negotiated proposals and in awarding construction contracts.

E. **The Construction Phase – Administration of the Construction Contract** generally includes:
    - Preparation of supplement drawings.
    - Review of the Contractor's Schedule of Values (a detailed cost breakdown of categories of materials and building trades); review of fabricators' and suppliers' shop drawings, material samples and equipment, and other required submissions.
    - General administration of the construction contract(s) including weekly visits to the site to review the progress and quality of work and to determine if work is proceeding in accordance with the contract documents.
    - Review of the Contractor's applications for payment, determination of amounts owing to the Contractor, and issuance of certificates for payments in such amounts.
    - Preparation of Change Orders covering authorized changes in the work with prior consent of the Owner.
    - Determination of the date of substantial completion and final completion; receiving, reviewing, and forwarding to the Owner the specified written guarantees assembled by the Contractor, and issuance of the final certificate for payment.
    - Coordination of tasks and schedule of the Owner's Landscape Designer.

Work that is <u>not</u> a part of the Basic Scope of Services:

1. Landscape design.
2. Furniture selection.
3. Wall and window covering selection.

Compensation to the Architect for the Basic Scope of Services as described in Paragraphs A through E shall be:

**Ten Percent (10%) of Construction,
not to exceed
Forty-Two Thousand Dollars
($42,000.00)**

February 29, 2000
Joe and Mary Client
Somewhere, US 12345
Page Three

Additional Provisions:

The Architect shall endeavor to design the project within a budget of $490,000, including landscape allowance and professional fees. In the event construction bids extend the project cost over $490,000, unless an adjustment in the project cost has been previously approved by the Owner, the Architect shall make appropriate revisions with no additional fees to the Owner.

Additional Services beyond the above scope, such as changes made after a phase has been completed or services beyond the above scope as described in Paragraphs A through E be invoiced on an hourly basis as follows:

Principal:	$130.00/Hour
Project Manager:	$ 95.00/Hour
Architect:	$ 75.00/Hour

The architectural fee shall be apportioned as follows:

Schematic Design:	20%
Design Development:	15%
Construction Documents:	40%
Bidding and Negotiation:	5%
Construction Administration:	20%
	100%

All blueprinting, photocopying, messenger service, overnight delivery, and postage shall be reimbursable to the Architect at 1.1 times the direct cost.

Invoices shall be submitted on a monthly basis, and are due within 30 days of invoice date. Payments beyond 30 days shall be subject to 1 ½% interest per month.

Upon acceptance of this proposal, both parties shall enter into a written agreement using the American Institute of Architects B151 Standard Form of Agreement Between Owner and Architect, 1997 Edition.

A $2,000.00 retainer is required at the acceptance of this proposal to be applied against the last payment to the Architect.

Respectfully submitted,

RICHARD PREVES & ASSOCIATES, P.C.

Richard Preves, A.I.A.
President

Accepted by: _____

For: _____

Date: _____

Builders' proposals need to be even more specific. Besides describing the architectural design service, all aspects of the project, from size, style, amenities, finishes and fixtures must be included. The only time you don't need these details is when there is an existing model of a new house, which sets the standard.

General contractors' proposals usually are tied to the scope of architectural drawings and specifications. As we will discuss in chapter 11, contractors' proposals must be carefully examined; often many allowances, exclusions, and substitutions are listed.

• A firm price or fee structure

Proposals should fully inform you of the fees associated with the services provided. We have previously discussed in chapter 3 how architects set their fees. Analyzing builders' and contractors' fees will be presented in chapter 11. Make sure that fee structures have a guaranteed maximum price and are not open ended.

• Terms and conditions of payment

Building professionals expect to be paid at different intervals. Make sure you understand the amounts and the schedule of payments required by each proposal. As you may recall, some professionals may try to obtain more money than is warranted during the first part of the project. Others will require retainers and deposits that should first be carefully analyzed. Before you sign any document, become comfortable with these contractual terms. If you accept a proposal, you are stuck abiding by its conditions.

• Offer and acceptance by all parties

A proposal will not be binding if it is not signed and dated by all parties. If you receive an unsigned proposal, it may not be a serious offer for services. On the other hand, if you sign and return a proposal, you have made a deal!

Refer again to figure 5-1 (see page 51). Notice the last paragraph contains a phrase I like to use, explaining that if this proposal is accepted, both parties shall enter into a more comprehensive contract. This acknowledges that this proposal is primarily a basis for a future, more comprehensive legal agreement.

Proposals can be very useful in first analyzing competing professionals and then choosing the best-qualified firm to do the work. Unfortunately, proposals are often the only legal agreements in force for many projects. Except for very small jobs, substituting a proposal for a contract can put the homeowner in a dangerous position. Even a well-written proposal will not contain important clauses essential to protect your interests. Don't start your building project without a good contract.

Contracts: The Good, the Bad and the Ugly

Proposals are not a substitute for contracts. Too many clauses are necessary to be included within the limits of a proposal. Two types of contracts are commonly used throughout the industry: *standardized* and *customized*. Custom contracts are usually prepared by each professional's legal representative. Often these documents are written to favor the firm presenting the contract, placing you at a legal disadvantage.

Let me share a brief story to illustrate this principle. A young couple who received a building lot as a wedding gift accepted a contract from a local builder to construct their house. They envisioned a brick ranch with oak floors, wood cabinets, and a marble master bath. But when they met with the builder's architect for the first planning session, they learned their budget was too limited for these amenities. They had signed the builder's customary two-page contract for a ranch house, which included none of the specifics they had discussed. Since they signed before any plans were drawn, and the brief contract included no specifications, the builder had no obligation to give them what they expected for the price they agreed to pay. Protect yourself by carefully reviewing any custom contract before you sign. First, let's discuss standardized contracts.

Standard Contracts Can Help

Professional associations such as the American Institute of Architects (AIA) have standardized architectural and construction contracts that have been prepared by attorneys and tested in court. These contracts are carefully written and are fair to all parties. The AIA publishes a comprehensive set of contracts specifically intended for use in smaller projects such as houses. For the Traditional System, two contracts are utilized for agreements with the architect and general contractor: AIA Document B155, *Standard Form of Agreement Between Owner and Architect–1993 Small Projects Edition* shown in figure 5-2 located in the appendix, and AIA Document A105, *Standard Form of Agreement Between Owner and Contractor–For a Small Project–1993 Small Projects Edition* shown in figure 5-3 located in the appendix. Each contract refers to a separate document that establishes the ground rules for the project, AIA Document A205, *General Conditions of the Contract for Construction of a Small Project–1993 Edition*, shown in figure 5-4 also located in the appendix.

According to the AIA, these contracts were developed so that "within a document family, common definitions and parallel phrasing combine to form a consistent structure in support of all the major contractual relationships on a construction project." Both the architect and contractor agreements adopt the terms and conditions of the A205 document.

As you see from these examples, blank spaces are available to fill in pertinent project data, followed by boilerplate, standard sections setting the conditions of the contract. I have found these contracts to be very useful, as most building professionals are familiar with their content. Many attorneys use them for a base contract and modify certain sections as they deem necessary. A great way to limit your legal

costs is to utilize these contracts with no modifications, if acceptable to your building professional.

Usually no one refers to a contract after signing unless there's a dispute. Then it's read with a microscope to review the requirements on how to settle a problem. If the dispute leads to litigation, decisions and remedies are based on the contents of the contract. The following is an explanation of some of the contract clauses with architects, general contractors, and builders. Some require a little interpretation for your understanding. My discussions are not meant to be a law school course in contracts. These clauses illustrate the importance of a written contract.

Contract with the Architect
(Figures 5-2 and 5-4, located in the appendix)

The contract between the owner and architect defines each participant's responsibilities, the specific work to be performed, and arrangements for payment. This is the easy part to understand. The following clauses require more of your attention:

• Document B155, Article 1 and Document A205, Article 4

During contract administration, the architect is only observing construction to advise you if the contractor is meeting the required obligations in accordance with the drawings and specifications. Litigation has established that the architect is not responsible for inspection or guarantee of the construction. This always comes as a surprise to most homeowners, as they picture the architect supervising construction at the site.

• Document B155, Article 2

This article requires you to furnish certain documents relating to your lot. Surveys describing the legal boundaries of the site, and engineering evaluations for the foundation if the soil characteristics are considered questionable, are two of the most common requirements. Plan to spend several hundred dollars to cover your part of the contract responsibilities listed in this section.

Article 2 states the owner must hire a contractor to build the project and provide the cost of construction. Architects do not guarantee the cost of construction, since they are not general contractors. They furnish construction estimates to the best of their ability. If your architect is a poor estimator, you may go through the entire design process only to have the construction bids come in over estimate. The architect could charge you an additional fee to scale back the design to meet the original budget. To protect yourself, consider adding a clause into the contract that requires the architect, with no additional fee, to redesign the project to reach the agreed budget if his or her estimate is off by more than 20 percent.

• Document B155, Article 6

The contract also details payments to the architect for his services, including reimbursable expenses. It is normal to reimburse architects for blueprinting and other indirect expenses above and beyond the basic architectural fee. You can, however, negotiate any of these points with the architect to limit your expenses.

Article 6.5 discusses revisions to the design. If you require the architect to make changes in the project's scope, quality or budget, you are required to pay additional fees. This is a very broad statement which should be further clarified by the architect. What constitutes a revision? Would moving a door on the plans require additional payment? Most architects anticipate changes during the design process and incorporate them into their basic fee. I recommend that prior to concluding negotiations of the terms of the contract, you have the architect spell out in writing exactly what will trigger additional fees.

You will notice that the contract does not address the issue of the architect's errors and omissions insurance, which protects an architect and his client from design errors. Because this insurance is costly, not all architects carry it. If you want this coverage for your project, make sure a clause is added to the contract.

Contract with the Contractor

(Figures 5-3 and 5-4, located in the appendix)

The construction contract begins by identifying the owner, contractor, architect, and the nature of the project. It describes the responsibilities of each party and defines the scope of work, usually by referring to project drawings and specifications. The contract usually establishes construction start and completion dates.

• Document A205, Article 5

The contract should include a procedure for change orders (work beyond the original scope of the contract) that requires a written description of the new work, with a written dollar amount. The change order should be signed by all parties before the construction change is made at the site. Changes often involve extra costs for the contractor. He passes these costs on to you, but predetermining limits can protect you from exorbitant charges. It's a good idea to fix the contractor's markup on changes to a maximum of 10 percent to 15 percent. See chapter 11 on construction change orders for more details.

• Document A205, Article 3.5 and 9

These two articles are combined to provide a comprehensive warranty. All materials and workmanship are covered for one year. After you finish chapter 12, which is dedicated to warranties, you will appreciate the beauty and simplicity of this comprehensive statement.

• Document A205, Article 7

The contractor will request payment every month, or when construction schedule milestones are reached. It is important that this schedule be established to avoid unforeseen financial obligations such as unanticipated interim payments. Milestones could be the completion of rough carpentry framing, drywall installation, or completion of the entire project.

Contractors usually require a maximum elapse time between payment request and payment, but it should be no less than fifteen days. If you use a title company to disperse funds, this time limit may be too short when the title company is busy. Have the title company specify the length of the payout procedure in writing.

• Document A105, Article 5

Specific insurance coverages are itemized within this article. Note that besides the contractor's requirements, the owner is also required to provide liability and property insurance. This can be accomplished through a standard homeowner's policy. The amounts and limits of each policy coverage should be included in this article.

• Document A205, Article 3.12

The construction contract should include clauses to protect you, the owner, from bad workmanship, faulty materials, and frivolous lawsuits. Included is an indemnification clause that holds you harmless for contractor negligence. This clause may protect you if litigation results from an accident that injures a worker or damages adjacent properties.

These are just a few interesting highlights of the standardized agreements between you, your architect, and contractor. Although they may seem long and cumbersome, they are actually the shortest the AIA has available. Much longer and more comprehensive contracts are also available to you and your attorney. For architectural agreements, AIA Document B151–1997, *Abbreviated Standard Form of Agreement Between Owner and Architect* is an eighteen-page document. Even more comprehensive is AIA Document B141–1997–*Standard Form of Agreement Between Owner and Architect*. The AIA also addresses construction contracts in a similar manner. AIA contracts A101, A107, and A201 are much more comprehensive, and require considerable time to comprehend. A few key concepts from these larger contracts, commonly used on commercial construction projects, could be added to their shorter cousins to your advantage.

For example, today's buzzword for settling disputes out of court is *Alternative Dispute Resolution (ADR)*.

Provisions appear in the larger contracts for resolving claims first through mediation, which is a non-binding means of using an independent third party to reach consensus. If mediation is unsuccessful, *binding arbitration* is the next step, using the rules of the American Arbitration Association.

Depending on whom you talk with, some attorneys prefer the regular court system. Mediation and arbitration can limit your rights and actually cost more than using the judicial system. In any event, adding a dispute resolution clause to our set of AIA contracts may be a worthwhile idea.

Design/Build Contracts

Owners wishing to use the Design/Build System can also utilize an AIA standardized contract, AIA Document A191–*Standard Form of Agreements Between Owner and Design/Builder–1996 Edition*, shown in figure 5-5 located in the appendix. This document contains two separate agreements. Preliminary design and budgeting are covered in part 1. Part 2 includes the services for final design and construction. If for any reason the project does not move beyond preliminary design, part 2 is not required.

Many of the same terms and conditions contained in the separate architectural and contractor contracts are also present in A191. These dual responsibilities are combined into the role of the Design/Builder. Reviewing this document, you will notice it is more comprehensive then A105, A205 and B155. Don't be intimidated by its length, and don't let a building professional convince you that these comprehensive provisions are unnecessary. Although these contracts may require additional work on your part, I know many projects where both homeowners and building professionals benefited from their protection.

All AIA contracts are available to the general public for a modest fee ($10–$15) and are available at each local AIA office. Check your yellow pages or call the national AIA office in Washington, D.C., (202-626-7300) for more information. Now that we have covered standardized contracts, let's look at customized contracts.

Beware of Customized Contracts

Customized contracts offered by certain building professionals require a more careful analysis than a standardized contract provided by a professional organization. Many of the disaster stories in this book originate from failure to analyze a customized contract and get legal advice before signing. Your goal is to have a comprehensive contract that is fair to all parties. Have your attorney review any contract before you sign it, regardless of the size of your project. The following areas are the most common land mines found in customized contracts.

• Deposit Refund

Your rights to have your deposit returned for default by the architect or builder can be limited or eliminated. Your deposit should not be encumbered through no fault of your own.

• Compliance with the Design and Blueprints

Some contracts may only require the builder to substantially conform with the house plans and technical specifications. This gives the builder free rein to make changes without your consultation or approval. You should be entitled to the design choices you selected.

• Time to Complete the Project

The builder may include in the contract a tremendous amount of time to build your project. Trivial reasons are often listed as excuses for extending the completion date, in some cases for years. For example, the builder could use unsubstantiated reports of labor shortages as an excuse for a delay. Your aim is to limit the builder's construction schedule to a reasonable time for prompt possession.

• Hidden Costs

Many contracts have hidden costs passed along to the owner. They can require you to pay for the builder's risk insurance, building permits, title insurance, and transfer expenses, which can add up to

big dollars. You don't need costly surprises. The contractor should identify all project costs when he presents the contract.

• Warranties

This is a subject often targeted by custom contracts. Your rights provided by state law guaranteeing a good construction job could be eliminated or severely limited by flawed warranties. I've seen builder warranty programs that state it is the nature of construction to shrink, crack or break! A close examination of a builder's limited warranty is as important as examining the contract itself.

• Dispute Resolution

Contracts should clearly address the means of settling disputes. Some contracts can limit or eliminate your right to use the regular justice system, substituting arbitration or mediation that may put you at a disadvantage.

These problems are just the tip of the iceberg, but they should help you to understand the need for legal representation before signing a contract.

Who Decides Which Contract is Used?

The decision of whether to use a standard contract or a custom contract can become a major obstacle in selecting a building professional. Often homeowners will decide to hire a firm, only to have an unfair contract forced on them so far into the process that it is difficult to halt the project. At this point, the homeowners are so committed to a course of action they will usually accept contractual language that places them in an inferior legal position. You can avoid this situation by requesting a copy of the building professional's contract during the proposal analysis and interviewing phase. A quick review of the contract by an attorney could expose potential problem areas.

If contractual problems arise concerning a custom contract, discuss these issues with the firm and see if you can reach a compromise. If the firm refuses to change the problem clauses in its contract, suggest using a standardized contract, even if it requires modification. If you and the firm can't reach an agreement, consider eliminating them from consideration.

If something is done incorrectly on your project, or is not functioning as it should, a written contract can expedite resolution of the problem. Simply having a good contract in force can be enough of a threat to force a settlement without going to court. If you do land in court, the judge will want to know what your contract says, and won't have time for abstract discussions of right and wrong. If you learn only one thing from this book, let it be this: *Do not proceed without a written contract.*

Finally, please keep in mind that I am not an attorney, and this is not an exhaustive discussion of contracts. Since it's vital to have an effective contract to protect your rights and interests, discuss all of your options with your attorney.

Protect Yourself with Insurance

We mentioned insurance briefly in our survey of AIA contracts. Whether you sign a custom contract or standard contract for your project, insurance requirements are an important part of that agreement. Never assume that a general contractor or subcontractors carries adequate insurance. The contract should require the contractor and subs to carry Workers' Compensation coverage, general liability, and risk coverage with adequate minimum limits.

As the project owner, you're required to provide your own liability, fire, and theft insurance to protect yourself from liability for accidents and injuries. If a carpenter is injured on your project, for example, you may be taken to court. With personal injury attorneys advertising on television for accident victims to get all they're entitled to, chances are that everyone involved with your project, however blameless, will be sued. Insurance is the one step you can take to protect yourself.

As the owner of a construction project, you could even be held liable for an accident miles from your building site. It happened to the client of an architect I know. A tradesman who worked for the roofing contractor lost control of a truck carrying a load of shingles to the client's building site. He collided with a car, seriously injuring the driver. Because the driver of the truck worked at the site and was headed there when the accident happened, the homeowner was among those sued.

• Confirm all coverages of your building professionals with certificates issued by each insurance company, stating the types, amounts and expiration dates of policies. Since accidents can occur at any time, require these certificates *before construction begins*. A typical certificate is shown in figure 5-6. This certificate includes the name of the carrier, the insured, and should also include your name under *additional insured*. This is an important feature, as the contractor's insurance company must also cover you!

• Consider taking out an *umbrella policy* for the duration of the project. This is a form of liability insurance that will cover you for most circumstances during construction, boosting your short-term protection. Coverage is inexpensive and easy to add to your homeowner's policy.

Not only am I not an attorney, I am also not an insurance agent! Check with your agent to discuss the different types of risk and exposure. If your agent does not have the answers, find one who does. Your building professionals are also a good source of insurance requirements. They like to have as much protection on a project as possible, regardless of the source!

I hope I haven't put you to sleep discussing these related issues. Unfortunately, like the rest of the world, the construction industry is not perfect. The importance of contracts and insurance cannot be overemphasized. Spend the time and resources to protect yourself and your family.

Figure 5-6
Certificate of Insurance

Producer	Issue Date: 01/01/00
Perfect Insurance	**Companies Affording Coverage**

	Company Letter	A	All-Risk Insurance

Insured	Company Letter	B	All-Risk Insurance
ABC Contractors			
P.O. Box 1234	Company Letter	C	All-Risk Insurance
Goodtime, U.S.			
	Company Letter	D	Owner's Casualty Corp.

Coverages

This is to certify that the policies listed below have been issued to the above named insured for the policy dates indicated. The insurance provided by the policies described are subject to all terms, exclusions and conditions of each specific policy. Limits shown may have been reduced by previous claims.

	Type of Insurance	Policy Number	Policy Eff Date	Policy Exp Date	All Limits (In Thousands)	
A	General Liability	ABC123456	1/1/00	1/1/01	General Aggregate	1,000
	Commercial General Liability	X			Products - Comp/Ops Aggregate	1,000
	Claims Made				Personal & Advertising Injury	500
	Occurrence	X			Each Occurrence	500
	Owners & Contractors Protection				Fire Damage (Any one fire)	50
					Med. Expenses (Any one person)	5
B	Automobile Insurance	DEF123456	1/1/00	1/1/01	Combined Single Unit	500
	Any Auto	X			Bodily Injury (per person)	
	All Owned Autos				Bodily Injury (per accident)	
	Scheduled Autos				Property Damage	
	Hired Autos	X				
	Non-Owned Autos	X				
	Garage Liability					
C	Excess Liability	GHI123456	1/1/00	1/1/01	Each Occurrence	1,000
					Aggregate	1,000
	Other than Umbrella Form	X				
D	Worker's Compensation And Employer's Liability	ZYX123456	1/1/00	1/1/01	Statutory	
					500 Each Accident	
					500 Disease - Policy Limit	
					500 Disease Each Employee	
	Other					

Description of Specific Items

Certificate Holder	Cancellation
Mr. & Mrs. Joe Client 1234 Main St. Somewhere, US 12345	Prior to the expiration date, if any policy is cancelled, the issuing company shall mail within 30 days, written notice to the named certificate holder. Failure to mail any notice shall not be an obligation or liability of any kind upon the company, its agents or representatives.
	_____
	Signature
	Authorized Representative
	Perfect Insurance

Chapter Five Recap

- Residential construction projects require comprehensive contracts for your protection. Without a fair legal agreement, your rights could be limited.

- Proposals differ from contracts. They are useful when comparing competing professionals, but lack the comprehensive details of a contract.

- Any contract should be reviewed by your attorney prior to signing.

- Two types of contracts are commonly used, *standardized* and *customized*.

- Standard contracts are available from professional organizations such as the American Institute of Architects. They contain standardized sections that are usually fair to all parties. This is by far the most economical choice.

- Custom contracts are prepared by a building professional's attorney and are often written to favor the firm over the homeowner.

- Prior to hiring professionals, you should become familiar with the contract they wish to use. If their contract puts you at a disadvantage, consider requesting a different contract or remove the professional from further consideration.

- Proper insurance coverage is a must for the homeowner and building professional before construction starts.

- Consult your insurance agent for required coverages and minimum amounts.

- Confirm all insurance coverages with certificates of insurance.

Chapter 6

Good Communication Saves Time, Money and Stress

Communication has been mentioned several times in previous chapters. In this chapter, we will explore how information is created and how to communicate that information effectively to all team members. The data we wish to share are the decisions you make for your project. Reduced to a simple level, you will accomplish your project by hiring a group of building professionals who must efficiently understand every facet of your program. Throughout our six-step process, communication between the participants occurs on many different levels, including oral, written, and graphic exchanges. Good communication is the lubricant to insure success for every activity.

I was once told that more than 2,000 decisions are required to design and construct a new residential project. I have never bothered to verify this number, but it is true that dozens of people will be making determinations that steer the course of your project. The building industry is a very interrelated field. The decisions of one party affect the actions of the other members of the project team. I know from experience, the higher the level of effective communication, the better the product. Here is a sequential list of communication combinations that occur during the process:

For the most part, homeowners are couples. They must be able to communicate their thoughts and ideas for the project. The dynamics of each relationship affects their ability to work well together. Sooner or later, they must speak with one voice to convey their final decisions. The methods used to reach this consensus varies; I am constantly amazed how often one half of the couple will make decisions without ever telling his or her partner. If agreement is difficult to achieve, everyone is in for a tough time. It is far better for the homeowners to speak with one voice instead of two. Effective communication certainly starts with you and your partner.

I recall a couple I worked with on a high-end addition and remodeling project. The process was a little rocky. Loud disagreements in determining the design were common. They had a difficult time working together and finding ways to reach decisions. Fortunately, regardless of a couple's relationship, the design is somehow always completed, and I breathe a sigh of relief when construction starts. But on this particular project, the problems got worse instead of better. Nearly every morning I would walk into my office to find a fax from the couple requesting changes to the design. I would update the drawings according to their revisions and forward them on to the contractor.

About six weeks into construction, we held a meeting at the site to discuss the progress and coordinate a few issues. When the general presented the additional costs for the changes, the husband seemed surprised, saying "What changes? I don't know anything about this. What are you talking about?" As it turned out, his wife had been composing the faxes by herself, without the knowledge of her partner. This was her way of getting those items she wanted, items her husband had ruled out during design! I made the mistake of assuming the fax instructions were common knowledge. Guess who bore the brunt of his wrath? If it was his wife, it didn't happen during the meeting. I was the one who took the heat!

There are countless areas of miscommunication between a couple that can cause a project to go astray. You must realize that it is very important to act jointly in making all decisions. Our six-step process is worthless unless the homeowners are unified in their efforts. Even single homeowners must know their own minds and make effective decisions.

Once this synthesis of ideas has occurred, the homeowners must communicate their goals and requirements to their building professionals. Now a third party has been added to the mix. You must be able to transmit your ideas so a complete stranger can understand what he is supposed to accomplish. Here is another potential stumbling block. Since homeowners are created differently, some will be better communicators than others. Even worse, a few have nothing of value to offer! Building professionals are not mind readers, if they cannot understand what you want, they have to use their best judgement.

Good communications are not necessarily related to the level of project expectations. Obviously, if a homeowner has a large budget with lots of goodies to choose from, communication may seem easier. Those with tighter budgets may have a lot less to talk about. It is often more difficult to produce good design on a tight budget because there is less to work with. But I have found that the opposite can also be true. Homeowners with the luxury of choice often overload and miss the forest for the trees. In their attempt to select every possible amenity, they overlook the importance of a central concept for living. On the other hand, homeowners with limited resources actually have to communicate more effectively, because without the amenities, the living concept must carry the entire design.

Here's an example of what can happen when a homeowner transmits the wrong information. A couple who had an unlimited budget became focused on big-ticket items. They had to have the latest in home entertainment systems, electric controls and every plumbing device in the master bath. A large part of a mountain departed the earth for all the granite that was used. Custom cabinetry and the latest in kitchen appliances abounded. But the overall design lacked a soul. The new house had no particular overall style. It was just an enclosure for all these expensive showy items. These homeowners failed to communicate their innermost thoughts to set the tone for their living environment.

Conversely, a couple with limited funds became so humble they failed to communicate their program. Assuming that they could afford very little, they did not take the time to explore the advantages of living simply. Great living environments do not have to be flashy. The Japanese have made designing modest, traditional houses into an art form. They chose to enhance a few basic elements in the design, transforming everyday elements like walls, doors, and floors into a singular concept of beauty.

Communicating effectively requires that you stretch your thought process to the maximum. If you are lucky to have a good designer on your project, take the time to explore the possibilities by encouraging an exchange of ideas. Good communication is a collaborative effort among the members of the project team. If you set ground rules that encourage creative thinking, all participants will respond by attempting to bring out as many ideas as possible. No idea should be too dumb to discuss!

How Drawings Communicate

Once you have conveyed your ideas to the project team, the designer takes the next step by translating your ideas into architectural solutions. Here the medium of expression changes. All those words and conversations you had with your design professional are now contained in the language of drawings. Since many homeowners have little experience in understanding these flat representations of a three-dimensional house, the architect must be able to effectively explain the proposed design. A good deal of verbal description by the designer is often required to transfer these technical drawings into images their client can understand. As the design develops and more decisions are made, the designer must assure the homeowner that the required information reflecting every choice is contained within the drawings. This can be a long process, accomplished during the course of several design meetings. Chapter 8 devotes a section to discussing the best way to communicate this information. Efficient communication is essential to ensure that you get the project you envisioned—and paid for.

Here's what can happen if homeowners fail to see the importance of communication with the person who designs their new home. A couple hired a builder who had worked for several friends of theirs. Communication problems started almost immediately when the builder told this couple he would attend all design meetings without his architect. "Architects talk too much and just confuse everyone," he said. "I know what you want and the three of us can work it out."

So the homeowners transmitted their goals and requirements to the builder over the course of a two-hour meeting. The builder then met with his architect, relaying the project data as he saw fit. The architect, without the benefit of contact with the homeowners, relied solely on the builder's information for his design formulation. This is an unfortunate example of the Design/Build System taken to an extreme.

When the builder returned with the first set of designs, the homeowners were surprised to see that only some of their ideas were included in the proposal. "This is not exactly what we talked about," they said. "Where are all the elements we wanted?" The builder replied, "I've been doing houses for a very long time. You really don't need a lot of those things, just trust me. You'll like what I give you!" In this case, communication was virtually nonexistent. The builder had appointed himself homeowner and architect and took over the entire process. He had built the same product over the years and his clients were given the choice of taking it or leaving it.

If design is a priority for you, and your project is more complex, expect to spend more time communicating during an extended design phase. On the other hand, if you have simple tastes, your design should zip right along. I believe homeowners want to build a custom designed house so they can live in an environment that reflects their personalities not available in tract houses.

The next stage of project communications does not involve you directly. All the decisions you have made must now be conveyed to everyone involved in constructing the project. Not only does this include your design decisions, but the hundreds of other technical details that produce a project. The architect must transmit specific requirements to the contractor, subcontractors, and suppliers. These are the basics, like how the house stands up, what keeps the weather out, and how utilities flow, to name just a few.

The medium for relating all this information is drawings and written specifications. This is how building professionals communicate. At this level, you are dependent on the project team you assembled. The more proficient they are in providing comprehensive documents and accurately reading their content, the more likely your project will be successful. Chapter 11 will inform you specifically what these documents should contain. Jumping ahead just a bit, these documents must include almost all of your project requirements. If the documents fail to communicate any specifics, there is a good chance you won't end up with the project you wanted.

These requirements are transmitted via a very technical level of communication that sounds like a foreign language. A similar example for the homeowner might be a diplomat whose very specific instructions must be exactly translated. As he listens to the translator speaking in the unfamiliar language, he wonders if every word and nuance of his intentions are really being faithfully and accurately conveyed. Just as he is depending on the skill of the translator, so too are you relying on the expertise of your project team.

Communicating with Secondary Team Members

The final aspect of project communications involves secondary team members, such as attorneys, insurance agents, financial institutions, and building department officials. This group handles the legalities and other obligations for the project. Their input is very important and can only be effective if they are given accurate information. Attorneys must understand the nature of the project and the project team so contracts may be properly prepared. Insurance agents need to be familiar with the specific risks of the project so adequate coverage is in place for all participants.

Government officials transmit many requirements at frequent stages of the project. I have seen several projects, especially additions, go astray because either the building department was not consulted, or their instructions were poorly communicated among the building professionals. Although we don't spend a great deal of time discussing these secondary players, their role is crucial to the success of your project.

You now can see that a project involves a large cast of role players who must be reading from the same script. Let me tell you a story of how good communications can help you through the potential pitfalls. Working with their builder, a homeowner selected the color of the plumbing fixtures in the master bath, deciding on white. Several weeks later, uncomfortable with their ability to coordinate colors for their new house, they hired an interior designer.

After reviewing their choices, the designer recommended changing several color selections, including the master bath. The designer documented the color change for the fixtures from the original white to hot pink in a memo that was sent to all project team members. Months later, the homeowners were walking through the nearly finished construction when they noticed white plumbing fixtures had been installed. When they questioned the builder why the fixtures were not hot pink, he responded that he understood the fixtures were always white. If they wished to modify the color, he should have been notified. Changing the fixtures at this stage would result in an additional charge.

Returning the next day with the designer's memo, they confronted the builder with the document, pointing out that he must have received it because all the other color changes had been installed correctly. Armed with written proof, the homeowner found it easy to convince the builder that he must foot the bill for replacing the fixtures.

Why You Need to Take Notes

Decisions made on a project usually are first formulated verbally, but the process falls apart when each decision is not documented in writing. In my experience, there are several key points during a project where specific written documentation is required. First, I recommend a detailed set of minutes be prepared for every meeting. Refer to figure 6–1 (page 68) for an example, *Minutes of Meeting*. These particular minutes are from a design development meeting between the architect and the homeowners, when they discussed revisions to the preliminary plans and details. The format is informal; each room that received attention is listed with the specific final decisions. You will notice that final decisions, pending decisions and actions to be taken are shown. Notes include the date and a list of all meeting participants.

At the beginning of a project, one person should be appointed to take notes, prepare the minutes, and circulate them to all project team members. If someone can't attend, make sure they receive a copy. On my own projects, I prefer to prepare these documents, so I am not dependent upon someone else taking forever to circulate them. If your builder or architect takes on this responsibility, you may want to jot down your own notes to double check the minutes when they arrive. If you disagree with any of the content, contact the professional and clear up the confusion! Meeting minutes can only be effective if they are accurate. Review them promptly and verify that corrections are made in writing and distributed to all. Depending upon the project size, you may accumulate minutes from just a few meetings or as many as dozens.

Another important document records the content of telephone conversations. Many discussions and decisions occur during telephone conferences, especially when the plans are being finalized. These exchanges are just as important as meetings, but easier to forget. Figure 6–2, *Telephone Memorandum*, is a useful format to document decisions reached over the phone. It lists the telephone conference participants and date of the call. This particular memo example addresses decisions made for a bathroom. The memo should be distributed to all project members, just like the minutes of meeting.

I consider telephone memos even more important than meeting minutes as miscommunication over the

phone happens all the time. It is very easy to misunderstand names of colors, model numbers and other specifics when you are not sitting face-to-face with your professionals. Homeowners also have varying degrees of proficiency when it comes to talking about technical issues. Ordinary telephone calls do not have the benefit of visual aids to help everyone understand what is being discussed. I recommend that all telephone call decisions be reconfirmed at the next meeting to avoid any confusion. E-mail is certainly a great way to document discussions.

Especially in today's world of voice mail, phone messages can easily be mislaid or misdirected. The telephone is a necessary convenience, but not a means to an end. I know a contractor whose voice mail message reads, "You have reached my voice mail, which is only a means of communication. If you would like to speak to me personally. . ." He has seen too many mistakes using voice mail. He understands that your 10:00 P.M. message on Saturday night could easily be misunderstood on Monday morning!

Throughout the remaining chapters, additional forms will be introduced to promote more effective communications. Obviously, this amount of documentation takes time. Some building professionals will bristle at pushing this amount of paper. They may say, "This is only done on big jobs. I can keep track of everything right up here," pointing to their head. Remember, any size project can have miscommunication. It only takes one error to cause headaches. Insist on documentation, even if you have to do it yourself.

Written communication not only helps the homeowner, but also aids the professionals. In my own practice, I find these documents worthwhile for my own peace of mind. First, it helps me to do a better job by accumulating a checklist to ensure I include everything that was discussed. As the drawings are nearing completion, they provide a means of double-checking my drawings one last time. Also, just like my clients, I need the protection. When I work with

Richard Preves & Associates, P.C.

Architecture   Planning

Figure 6-1

Minutes of Meeting

Client Residence
December 13, 2000

1. The revised asymmetric lower level was selected. The north wall of the East Bedroom Suite will be extended approximately 2' into the Storage Room.
2. The East Bedroom Bath door will be relocated.
3. The West Bedroom Closet door will be relocated.
4. The Lower Level Wet Bar Cabinet could be furniture instead of millwork cabinetry. To be determined.
5. The east Dining Room wall at the south wall return will extend up to the ceiling. The Pantry and Closet will have a ceiling at 8'.
6. An exhaust fan will be installed for the Kitchen. The louver shall be above the east wall cabinets. The exterior vent termination shall be on the vertical wall above the Garage.
7. An exterior keypad will be used for the Garage doors.
8. Two skylights shall be installed in the Guest Room, south side.
9. Two "outlook" benches shall be added to the screen porch walkway.
10. Lighting design was discussed. No downlights shall be used in the Study or the Sun room.
11. Floor finishes for the Garage, Hall and Laundry Room shall be sheet vinyl or vinyl squares. Floor finish for the Hallway between the Kitchen and the Utility Room shall be wood.
12. The Great Room wood floor will run parallel to the north/south walls with no pattern. The wood will be extended from the Entry and Hall into the Powder Room and Shower.
13. The stair design was discussed. The Architect will price-up solid stone slabs versus solid wood treads. The Architect will also provide sketches and/or pictures of steel railings.
14. The fireplace will have no border around the firebox opening. The hearth shall be stone, flush with floor level.
15. The Interior Designer will send the Architect an updated furniture plan for lighting design.
16. The Architect will send Joe 2 sets of updated floor plans for security and audio planning.

End Of Minutes

Richard Preves & Associates, P.C.

Architecture   Planning

Figure 6-2
Telephone Memorandum

The following is a summary of a telephone conversation regarding this project.

To: Joe Client

From: Richard Preves

Date: May 1, 2000

Subject: Client Residence

Hall Bath

- Villeroy & Boche™ Sinks
- Corian™ or like top countertops
- Wood – (stained finish) Cabinetry
- Kohler™ – white water closet
- American Olean™ 6" x 6" matte tile "Snow Mist"

Entry

- Emerald pearl granite with white key inserts for floors.

homeowners who constantly change their minds and then conveniently have "selective memory," I use the documentation as my back-up. That's why I take my own notes—turnabout is fair play!

Good record keeping therefore benefits everyone. Just as I have files organized in my office for each project, I recommend the homeowner do the same. You will be surprised by project's end how much paper will be produced. Group your files according to the different phases of the project; the organization will pay big dividends down the road.

Don't Overlook Stress

While we are discussing effective communications, I would like to consider a topic rarely mentioned in residential construction books—stress. Building projects often produce tension between family members, create strain or amplify existing anxieties. When I start working with a couple, I usually try to introduce the subject in a humorous way. According to an article in a psychology magazine, the second leading cause of divorce, right behind infidelity, is undertaking a building project. While I'm not sure about the statistics, I often find myself smack in the middle of the dynamics of a marriage.

Some relationships are pleasant, others have been stormy at best. I have sat through arguments, screaming matches and insults resulting in the abrupt termination of meetings. Occasionally, I have had to separate arguing couples, sending them to separate rooms to cool off. In college, we were required to take an introductory psychology course. At that time, I had no idea of its future value! If you are experiencing stress in your relationship with your partner, I guarantee a construction project will only serve to increase the opportunities for conflict.

One of my more stressful projects involved designing a very expensive new house. I was attending a meeting with the couple at the husband's office, trying to firm up some final design development material selections. He had to step out of the conference

room for about twenty minutes to take a telephone call and asked us to continue working. Up to this point in the process, meetings had often been laced with arguments and lines drawn in the sand when it came to making mutual decisions. Some meetings actually canceled out previous decisions instead of breaking new ground.

When the husband returned, he asked if we were making any progress. "You bet," his wife replied, "Richard threw me down on the conference table and made mad, passionate love to me!" It took several minutes for me to wipe the astonished look off my face, for the wife to stop smirking, and for the husband to cool down. I now knew that previous arguments over the location of a towel bar in the bathroom had nothing at all to do with the real issues in this marriage!

A prudent step in the project process is to identify possible points for disagreement and establish ways of dealing with them. I would like to introduce potential issues, which, in my experience, are perennial hot buttons. Good communication skills can assist couples in getting over these bumps in the road. Remember, the building professionals you hire are not therapists. They should not be the referees making the calls to keep the project on track!

Anticipate Major Stress Points

Aesthetic decisions often cause disagreements. When confronted with the multitude of decisions that must be made, two adults may discover that they have different tastes. I've learned that choices of styles, materials, and colors usually top the list when it comes to differences in opinion. Aesthetic thinking is a personality characteristic that is formulated and ingrained over the years. Its importance also varies for everyone. I find that most couples react to aesthetic issues in one of three ways. First, the lucky ones agree on almost everything, making the design process a breeze. Secondly, one member of the couple won't care and will defer to the other's selections. Whether this is through indifference or the dynamics of the marriage, I am not sure. Finally, and most unhappily, a couple has little in common and cannot agree even on the basics of design and materials.

We will skip the first two cases, since disagreement isn't involved. Resolving aesthetic differences is a lot like politics; it usually ends with a compromise. The good thing about styles, materials and colors is that there are so many to choose from. If one likes traditional and the other modern, blending styles is very common today. When my wife and I built our house, some of the rooms went unfurnished for years, because we couldn't find the pieces that we both could agreed on. The interior color of our house is nearly all white because that was the only color we both liked. Our walls were devoid of artwork for years, until we stumbled on something that appealed to both of us. We found that if you look long enough at all the possibilities, a compromise can usually emerge without causing any stress.

Today there are numerous choices, which should encourage compromise. Chapter 8 will address the issue of style, explaining why there are no longer any pure design styles. The majority of residential houses are really a mixture of many historic stylistic elements. Combining different styles today is very acceptable. Aesthetic compromise relies upon openness to new ideas. Let your designers show you all the possibilities. The project process is long enough that it affords the time to give proper consideration to each choice. In my own case, my wife and I compromised on simple, traditional antique wood pieces for tables, bookcases and buffets, and more contemporary furniture for couches and chairs. Her taste for pictures with landscapes and my preference for purer styles led us to antique travel posters of European outdoor settings. These compromises took years to evolve, but the delay didn't bother us. Our walls are still painted white!

Financial issues are another leading cause for bigtime stress. Couples often have differing ideas about the amount of money that should be spent on a project. We could talk at great length about the merits of either economizing for greater financial security or overextending by becoming "house poor" for the first several years. With mortgage programs allowing for low down payments and easing the rules for qualifications, house budgets have certainly increased in the recent past. My goal is for you to set a realistic budget, stick to it, and avoid cost overruns. If you need help in determining that bottom line number, financial advice abounds today.

A frequent problem with project finances is that a project budget is agreed upon, only to be exceeded by one or both partners. We have mentioned that adding one design "extra" every week can result in a substantial cost overrun. We also discussed how determining a project cost early on can be difficult when you are unsure what your money will buy. Compromising on money is very similar to agreeing on aesthetics; eventually you will reach a number that is acceptable to both. Financial agreement is the most difficult, however, and should be established as early in the process as possible.

Let's use our car example to explain how you can agree on a workable construction budget. A common occurrence is to walk into a dealership already set on the car to buy, only to be tempted by either a bigger model or luxury options. By the time you drive out of the showroom, you probably spent more that you originally planned. Are you happy? That depends on whether you can afford the financial consequences. Chapter 9 discusses this tendency to overspend and is devoted completely to project budgets. If you can decide how much you can afford, I can help you bring in your project close to that number. But first, you and your partner need to agree on the amount.

Remodeling Can Bring Stress

Aside from aesthetics and money, another leading cause for stress is living through the inconvenience of an addition or remodeling project. Most homeowners don't have the luxury of moving out during this type of construction. They are stuck coping with the anxiety of daily interruption, dirt, and destruction to their home. If you intend to remain in your house for the duration, you will become best friends with construction workers for several months. They will greet you at 7:00 A.M. every morning, make loud noises, lots of mess, and occasionally disrupt the electricity and water.

Although exciting at first, the action and progress you have been anticipating for so long will soon change to frustration, as you lose a working kitchen, bathroom, an entrance into your house, or anything else you can't live without. Planning such a project takes a lot of coordination to assure an alternative routine for daily living. Stress certainly plays a big part in such a project. Good communication and a dose of patience is about the only way to endure the siege. At the beginning of each addition or remodeling, I dedicate part of a design meeting to discussing just these issues. We talk about how the family will live, where all the furniture and personal property will be stored, and we try to develop a realistic schedule, room by room. If I sense that my clients do not handle stress particularly well, I address this issue head-on. I want to avoid any surprises and let everyone know what to expect!

I recommend a set of ground rules that minimize surprises be established to control construction scheduling. If the construction work week and hours are determined in advance, you won't meet a plumber in your bathroom at 6:00 A.M. Sunday. Milestone dates are also helpful: The kitchen will be dismantled on May 1st, and will be back in business June 1st. The garage will be used to store furniture and materials and will be unavailable for cars throughout construction. The master bath disappears May 15th and will reappear June 25th, and so on. If the water service will be shut off for two days while a new service is installed, a short stay at a motel is probably in order.

This coordination is only accomplished with careful planning involving thorough communication among all members of the project team. Do not start this type of project without first knowing what you will experience. Many think they can weather the storm, only to be overwhelmed by the stress, resulting in family turmoil. If the project schedule starts to fall behind, even better communication skills will be required. Plan on having some meals at the local eatery and several family diversions away from the house. If you can afford it, a vacation is the best!

Another issue that affects additions and remodelings is safety. Children are fascinated by construction and often can expose themselves to dangerous conditions. Open construction provides all kinds of opportunities for cuts and falls. Power tools left lying around are dangerous targets for exploration and experimentation by curious children. You must carefully communicate with them what is going to occur and organize a set of rules they can understand before construction begins. In this case, an ounce of prevention is worth a pound of cure!

Inquisitive children are also famous for distracting the tradesmen. Questions and requests for nails and pieces of wood are cute once in a while, but your project will slow down if children are allowed free rein in the construction zone. Since kids will also be upset by changes in their routine and environment, it is important to balance their need to participate in the process with their safety.

Managing Stress

To summarize, if your family is currently experiencing stress, a project will only serve to magnify those issues. We have identified several points for high anxiety, which require effective communication to help you through the process. From my experience of both constructing and living through these projects, I can suggest the following steps:

• Identify probable areas for disagreement and establish ways of dealing with them in advance. If couples know themselves and the dynamics of their relationship, these hot buttons should be fairly easy to anticipate. No matter how well you manage your project, stress cannot be totally eliminated. Just as you discuss your design program, you can also anticipate stressful events on the horizon.

• Aesthetic and amenity issues are tough nuts to crack. Compromise here is the name of the game. Try reaching an agreement by using "trade-offs." For example, "I get the three-car garage, if you get a sewing room," or "We buy the home entertainment surround sound system for me and the hot pink bathroom for you." Build on whatever common ground you can find to solve any remaining problems.

• Agree as soon as possible on a budget and stick to it. Get good advice to verify your financial projections and assumptions. Be ready to compromise to obtain the best value for the money you spend.

One final point about managing your project through good communications: Being in control of your project does not give you the license to be a neurotic dictator. I once had a remodeling and addition project where I had given a set of my newspaper column articles to my clients to review and familiarize themselves with my approach. Unfortunately, as the project proceeded, they took the communications article a little too seriously. On a daily basis, the general contractor and I were inundated with three to four phone calls and as many faxes. Often the issues were repetitive, with the same answers constantly being supplied. About halfway through construction, all this began to resemble the little boy who cried wolf. The calls became disruptive and ineffectual.

If you hire good professionals, your input is certainly required, but give them the room to do their jobs. Call or issue a communication only when you encounter an issue or a question that hasn't been previously answered in depth. *New House/More House* is about controlled, efficient project management. I encourage professional conduct from all parties, including the homeowner—not a constant barrage of unproductive, repetitive, or irritating communications. The more comprehensive and well timed your communications, the more effective the result.

Chapter Six Recap

- **Communications occur on many different levels throughout the project process.**
  - **Homeowner couples exchanging ideas**
  - **Homeowners expressing their goals and requirements to their professionals**

- **Building professionals providing required information to all team members.**

- **Effective communications are a must for a successful project. Committing all communications in writing enhances the chances for a successful project.**

- **You and your building professionals should utilize several standardized forms in keeping track of all decisions.**
  - **Minutes of Meetings**
  - **Telephone Memos or E-mails**

- **Family stress is a by-product of residential construction projects. Several potential issues concerning aesthetics, money, and inconvenience are a few of these "hot buttons" that can cause family strife. Good communications and compromise are the best ways to deal with these problems.**

Part Two

What You Need to Do

Chapter 7

Take the First Step by Hiring a Building Team

Now that you have learned the basics in part 1, you are ready to tackle your own project by applying the six-step process. The following chapters are arranged in a sequential course of action for developing your project. Let's first begin with hiring the right professionals to accomplish your project.

I have a distant relative who accepted a new high-level job, which meant moving to a new city. To make the move an exciting experience and encourage a smooth transition, he decided to build a new house. Family members would play a role in planning their part. Although his teenage children were unhappy leaving their friends and school, designing their new rooms intrigued them and reduced their protests.

While still commuting between his old home and new position, he asked people in his new office to recommend reputable builders. Two of his co-workers mentioned the same company, saying they understood the builder developed a lot of new and interesting houses. My relative took their advice and met with the president of the firm. The president showed great enthusiasm for the project and presented an impressive portfolio of very lavish, newly completed homes.

Lacking any friends in town and working all day to meet the challenges of the new job, my relative accepted the president's dinner invitation to discuss working together. Impressed with the sincere efforts of the company owner, "Mr. Smith," to welcome him to the city, he hired him on the spot. "Feel free to drive past any of our projects, see if you can find one that appeals to you. Knock on the door. Everyone we work for is real friendly," he said. My relative, who I'm happy to say belongs to a distant branch of the family, didn't bother to do either. He was certain he could form a good working relationship with Mr. Smith.

Four months later, he found himself working instead with a young project manager and very little progress to show for the time spent and the large deposit paid. When he asked when the president was going to get more involved, he was told, "Mr. Smith is usually either selling projects, playing golf, or on his boat. He rarely works directly on each project, and of course, your home is one of the smaller projects in the office right now."

Finding and evaluating the right building professionals is one of the most important early steps in getting your project off to a good start. This chapter is devoted exclusively to locating potential candidates, asking the right questions and evaluating the information presented to you. I will explain specific issues to address and what a professional should provide for your consideration. At this stage, you need a very definitive plan to follow. Otherwise, you may join my relative in going down the wrong path.

Finding Building Professionals

Our first step, and possibly the most difficult, is locating good, prospective candidates. The great majority of professionals do not advertise, and they are probably not household names that immediately come to mind. Building professionals obtain work through word of mouth. They are dependent upon their reputation spreading through many channels, such as the cocktail party circuit. You must find these channels and tap them for whatever worthwhile information you can obtain. The following is a set of proven methods I recommend you try.

The most common course of action is networking with family, friends and business acquaintances. If you have a wide enough circle, you can gather all kinds of good references as well as being warned whom to avoid. References from individuals who worked directly with a company are the best type, since they have firsthand knowledge. Much of the information you will hear, however, is second hand and will be less accurate. You will be told a lot of stories similar to this one. "You know so-and-so, she is so picky that if anyone can work for her, they must be good." Another common one is, "Oh, all my friends have used XYZ Builders. I'm sure they're very qualified."

As you reel in this raw information, you must become a bit of a psychologist. Remember the old saying, "Consider the source." If a reference is coming from someone you feel has his head on his shoulders, it probably is of value. A referral from a flaky friend may be only as worthwhile as your opinion of this acquaintance. Consider each recommendation with good sense, regarding the source.

A second resource revolves around checking local business people who are involved with the building industry. Realtors are excellent sources for information, they know all the players in town. Their advice could produce good results, providing they have no direct business relationship with their referrals. People in your local building department can also be a good source. Although many prefer not to make recommendations because they are a public agency, some are willing to talk. They are in a great position to evaluate professionals; after all, they review the plans and inspect the construction. If you can loosen them up to speak, they could tell you who knows how to do it right, and who to avoid. Because they do not want to be accused of slandering someone's reputation, playing favorites, or taking liability for a bad reference, they may be reluctant to give advice. If this is the case, ask them to give their opinion of specific professionals you mention first.

Another industry related resource is local building material suppliers, such as lumberyards, home centers, and kitchen and bath remodeling stores. Just like Realtors, they can provide useful information as long as they don't favor their own customers. Their referrals are probably less reliable, however, because they sell the materials without seeing the finished product. However, they certainly will know who is doing the most construction volume.

If you like to drive around, another method is to visit neighborhoods where different types of construction projects are ongoing. Drive through older areas where additions and remodelings are more prevalent. Visit new developments to check out new house construction. Usually you can find a sign posted in the front yard announcing what company is doing the work. Can't find a sign? Politely walk over to the site and ask a tradesman who is involved with the project. This is an excellent means of connecting company names with the type and size of project they perform. Take your camera along, take pictures and keep a log with the pertinent data for later reference. Not only is this a good way to find professionals, it may also enable you to get some ideas for your project!

I consider the yellow pages your last resort to find potential building professionals. Cold calling each listing is a time consuming chore. You will repeat your inquiry several times before you reach the right person, only to be told that firm doesn't work on your type of project.

If the telephone book is your only resource, this screening process, although time consuming, will finally yield a short list of professionals you consider qualified. However, you will be gathering references from strangers instead of friends. I can remember only a few projects that came into my office through cold calls from the yellow pages. Referrals by word of mouth is really the primary means of finding potential practitioners in this business.

Once you've assembled a list of prospective candidates, you can research their qualifications. This will require meeting the professionals, reviewing their experience and portfolio of past projects to determine if they make the first cut.

Interviewing Prospective Professionals

Do you remember your last round of job interviews? You probably polished your resume, dressed for success, sat up straight in your chair and carefully considered every word you said. We are going to draw upon these experiences, except now you're on the other side of the table! You are now the employer, gauging the qualifications of each candidate you might entrust with a good deal of your money. Use your people skills to form an impression of each professional. By engaging in conversation and soliciting answers to your questions, decide whether you feel a trustful relationship could be established. Is this someone you could be friends with, or does he remind you of a used car salesman? Take notes of your opinions formed during several interviews for each final candidate.

To begin, the following interviewing techniques in this discussion apply to both the Traditional System for architects and the Design/Build System for builders. Although many of the points in this chapter will apply to general contractors, further interviewing information for this group of professionals is contained in chapter 10. Your goal is to gather as many referrals as possible from a wide variety of sources for each professional you're considering. This will help you form a picture of each candidate's experience and current activity. The more information you gather, the clearer the image will become.

I recommend interviewing several candidates from both project delivery systems. Even if you have already formed an opinion as to which system may be best for your project, exposing yourself to both possibilities is time well spent. If you prefer one delivery system over the other, you still may want to consider design/build instead of the traditional system, or vice versa, depending on where you find the best qualified professional. If I had to rank the importance of professional qualifications and project delivery systems, I would choose working with the best professional first. The amount of interviewing I suggest is going to take some time, requiring your patience and perseverance. If you are the impatient type and only have limited time to spend on your project, this is the place to slow down and take your time. I can guarantee that if you are dedicated to the interviewing process, you will increase your chances for success.

The location of these interviews plays an important part in forming your opinion. Although some professionals may prefer to meet you at your current residence, try to arrange at least one meeting at their office. One look around their place of business can tell you a great deal. You immediately can determine the size of their staff and the type of projects currently on the boards. Is their office neat and orderly? If the professional runs a tight ship, chances are your project will be well organized. In contrast, drawings and files lying helter-skelter could be a good indication of less than desirable performance.

I feel experience is the key to evaluating potential building professionals. First, professionals usually work on either residential or commercial projects, although a few do both. In addition, residential specialists may only work on projects of a certain size or style. Some architects, for example, will only take on new houses or significant additions, leaving smaller additions and remodelings to others. Certain contractors' main interest is building additions and remodelings, because they prefer short-term projects.

Your goal is to match your project's scope to the experience of the building professionals. This concept is of paramount importance. If you are planning a small addition to your home, do not attempt to hire an architect or builder whose forte is expensive new custom homes. Even if they would decide to accept the job, your project would probably receive the least amount of attention. By the same token, a professional accustomed only to small jobs stands a fair chance of being overwhelmed by a much larger project. You will be best served working with a firm whose "bread and butter" is your size project. There is no substitute for past experience. Just as you might hesitate to permit an inexperienced surgeon to give you a heart transplant, use the same judgment in selecting building professionals. Match their strong suit to your building program.

Many architects will only design in a certain style. They have spent a good deal of time perfecting their approach and may be reluctant to switch their technique. Even worse, they may insist that you adapt to their style, forcing you to settle for something other than what you really want. Builders can be very much the same. For those professionals who value the bottom line, constructing the same type of project is more profitable. Everyone on the job can think less, speeding up the process and minimizing mistakes. Although I hesitate to compare your match with a building professional to a marriage, in both cases, common values and characteristics will go a long way toward smoothing the bumps in the road.

Speaking about experience, return to your first job interview and remember your frustration about being qualified but rejected because of lack of previous experience. "How can I get experience without first getting a job?" you might have said. The same principal applies in the building industry; everyone has to start somewhere. Most young practitioners begin with smaller projects and work upward, earning their spurs. Older professionals who gained experience while working in other firms are ready for larger projects even if they are just starting their own companies. If you come across a potential candidate that excites you but lacks the background, you must decide whether to take a calculated risk. Everyone needs a break sooner or later.

As a young architect, I tried to make my own breaks on projects I dreamed of obtaining by offering either free preliminary design to show what I could do, or offering to bring in an older professional to act as a consultant to ease the fears of potential clients. After getting a few nice projects under my belt using these methods, I was ready to roll!

Every architect or builder should have a "portfolio" or a collection of pictures of completed projects. This is the best means in familiarizing yourself with each candidate's work. Homeowners have a tendency to quickly page through these pictures. Take your time and look closely at each project. Have the professional explain certain projects, what was special, challenges they overcame, or what they liked best. These photos reveal each firm's style, project size, project complexity and the number of completed commissions. As an author can point to books on a shelf, the portfolio is the evidence of a building professional's experience.

This discussion brings to mind a story of a homeowner I met years ago. He and his wife were building a new expensive custom home and were interviewing potential architects. They told me of their surprise and confusion when three different candidates had the same house in their portfolios! On the second interview they thought it may have been a coincidence, but at the third interview, they finally asked the question. The flustered architect explained that he and the other two professionals the homeowners named used to work together at the same large firm before they all started their own. Since they were all fairly young, they needed to show experience and depth in their portfolios, so they borrowed this project, rationalizing that since they had all worked on it, they could claim it as their own work. Actually, this is an acceptable practice as long as the professional acknowledges up front that he or she worked on a particular project while working for another firm. If a young aspirant displays many large projects, you may want to question where all this experience came from.

How to Check References

For each professional whose portfolio displayed projects that match your project's particulars, request a list of references from completed commissions. References for professionals are a necessity, similar to their portfolios. This list should be immediately available; reject any candidate who tells you it will be available in a few days. My referral sheet usually contains five to six names with addresses and telephone numbers. I try to provide a mixture of older and newer projects for a better sampling.

You have arrived at a critical point in the six-step process to insure success for your project. Many people are hesitant to call or visit complete strangers listed on a reference sheet. They prefer trusting their instincts about a certain professional rather than calling past clients. If you fall into this category, overcome your shyness, buck up your courage, and make the calls! The time you spend talking with other homeowners from previous projects will pay big dividends. Not only can they tell you their opinions of each professional, you will be surprised at the pointers they can provide. I know my list contains individuals who are happy to talk about their project experiences. After all, they were in the same position as you not too long ago! Checking references is probably the single most important task for homeowners on their way through the six steps to success. Do it!

Here's how you can begin your interviewing. Prior to making your calls, write out a short list of questions that you plan to ask and the information you need. Since it is easy to become sidetracked during these conversations, use this list to double-check that you have covered all the bases. See figure 7-1 (page 83) for a sample questionnaire.

After introducing yourself and explaining the purpose of your call, verify the professional's exact involvement in the project you've seen. What role did he play in the process? Was he a main player or just a peripheral participant? Determine exactly the services the candidate supplied; often you will find that some professionals exaggerate their contributions to a project. Reviewing chapter 3 for our discussion of different services offered by building professionals would be beneficial in forming your list. Check off each service provided by the candidate on the project. This will help you form a clearer picture of the professional's involvement.

Ask whether the professional delivered his proposed services as promised. At this point, you must listen with a degree of caution. It is my experience that many homeowners, even after the completion of their project, still do not understand what services the contract required their building professional to provide. Homeowners unfamiliar with the process may depend on stereotypes in forming their opinions. "That architect never came to the site to supervise construction, that's what architects do, right?" is a typical comment, even though that service was not part of the architectural contract. Another misconception goes something like this: "The general contractor never offered his help in selecting our interior finishes and colors." By now, you should know this is not the job of a general contractor, but the responsibility of the design professionals or builder.

Don't assume all homeowners are knowledgeable about the agreements they signed. I have had clients who constantly required repeated reminders of what services they signed up for. No matter how much I try to educate them, some don't get it. Try to verify whether services allegedly not performed were really included in a homeowner's contract, or whether the professional actually shirked his or her duties.

Calling references is fairly easy, as homeowners don't need much prompting to talk about their experience. Those with recently completed projects tend to have more to talk about, as their encounter is fresh in their minds. Older references are usually past the initial flush of excitement and only remember a few points. Regardless of which reference you call, remember one fact: Just like job-seekers, building professionals will list only their best projects. No one wants to give a disgruntled former client the opportunity to vent frustrations. If you hear only lukewarm responses from references you call, beware. There are sure to be several unhappy clients out there who were not listed. If candidates don't have enough satisfied clients and must list halfhearted references, consider rejecting them. The same reasoning applies to those professionals who only provide one or two referrals, unless they are just starting their companies.

The ultimate reference check is visiting homeowners in their completed house. During your telephone conversation, ask if they would be comfortable permitting you to briefly tour their home. These field trips will enable you to see firsthand the design, materials, and quality of workmanship that pictures can convey only marginally. Older projects will also reveal how they have withstood the test of time. Poorly constructed homes will suffer after many years with foundation cracks, drywall joint splits, and doors that no longer close.

Meeting the homeowner also provides an excellent opportunity to gain more information about a candidate. Past clients who had a positive experience generally are proud to show off their project. But remember, you are a guest in their home; let them lead you on the tour. Using your people skills, engage your host or hostess in conversation, politely working in your questions. Ask them if their professionals stood by their work, returning after completion to help with problems. Determine if their project met the goals of our six-step process. Did the finished product meet their expectations, come in on budget and finish on time? Finally, remember to ask the ultimate question—would they use this professional again?

Returning to your interviews with prospective professionals, here are more questions you should ask. Your first goal is to determine who exactly will be working on your project. Many residential architects and builders are small companies; the owners meeting with you are the people who will work on your project. However, if you are considering firms with many employees, find out who will actually be working with you on a daily basis. From my story at the beginning of this chapter you'll recall how my relative never saw the owner of the company he hired after the project began.

Owners of large companies are often famous for meeting with you until awarded the job, only to hand you off to some junior staffer. This is a variation of the old bait-and-switch routine. The difference in experience and skills among young professionals can be alarming! Part of your interview should include meeting the specific project staff, not just the owners. If you are comfortable with the owners, but have reservations about the staff you apparently must work with, try insisting on working only with the owners. Tell them you want to work only with the "A Team!"

The next question should determine the company's present and future workload. This can be tricky to pin down, as professionals are always uncertain about future work. The nature of the building industry requires professionals to submit proposals or bids on as many projects as possible because only a fraction of these projects will come their way. For the aver-

Figure 7-1
Reference Interview Form

Professional: _____

Reference Name: _____
Address: _____
Telephone Number: _____
Email Address: _____
Fax Number: _____

Type of Project: _____
Date Completed: _____
Services Provided: _____

- Site Analysis: _____
- Programming: _____
- Design: _____
- Finish Selections: _____
- Interior Design: _____
- Construction Documents: _____
- Estimates and Budgets: _____
- Financing: _____
- Construction: _____
- Construction Administration: _____
- Site Visits: _____
- Punch List: _____
- Warranty Work: _____

age professional, success may be measured in obtaining one project for every two to four proposals submitted. Occasionally, architects and builders will be more successful than the above average, receiving more work that they can manage. It is therefore difficult for a professional to always tell you their present and future workload because they simply cannot make an accurate prediction.

To make a determination, ask each firm about its current projects and the number of proposals still pending. By ascertaining the size of the professional's staff, you can make an educated guess as to whether the firm has more work than it can handle. A good rule of thumb is that each staff member can handle two to three projects in different phases of competition at any given time. You don't want to work with a company that is overextended. Your project won't receive the attention it deserves. During a hot economy when everyone is busy, your first several choices of qualified professionals may not be available unless you are prepared to wait.

An honest architect or builder will tell you he can't start your project for weeks or months. Less reputable professionals will try to string you along, making little progress until they finally have the time. If you are willing to wait for your first choice, fine. But if you feel you must start right away, you may have to settle on a lesser choice to produce your project. This is a tough call involving calculated risks.

Discuss with each prospect how he typically approaches a project such as yours. Have them describe in detail the services he offers. When I attend interviews, I show the homeowner similar past projects, stating the program requirements of the client and how my design solution fulfilled their goals. For example, a completed project may have presented a site almost too small to accommodate the project's size. The solution might involve taking advantage of large window areas to make each room appear larger. This conversation will divulge the architect's

or builder's creativity and attention to detail. He should describe to you the sequence of steps and methods he would use to produce your project. Practitioners who enjoy the process will be happy to talk your ear off describing their efforts. Those who view each project as just another job may be less vocal.

If your program requirements are far enough advanced at this point, ask how each professional pictures a possible solution. This puts the candidate on the spot, but is certainly a fair question to ask on additions and remodelings where alternatives are limited. Depending on each professional's thinking process, you will receive a variety of answers. Some may be noncommittal, others may have a good idea from the very start. Don't be discouraged if some professionals decline to offer much comment. I know that I take a large gamble if I volunteer an idea that doesn't appeal to a prospective client, placing myself at a disadvantage in obtaining the project. I usually try to make a few broad comments without committing myself. At this stage, you simply want to further the conversation with each professional to gain more information in forming your opinion.

Finally, ask each prospect about the company's history and present ownership. Determine how long it has been in business. Fledgling firms are easy to understand; they have a brief pedigree and you are probably talking to the originators. Older companies may have experienced changes over the years, including firm names and ownership. Find out if they have always operated under the same title, or was the company formerly known by another name. Trace the chain of ownership from the beginning to the present. Occasionally you may hear of homeowners checking out an established, well-known firm, only to find that the original owners whose reputation they were seeking either sold their interest or retired, leaving new untested owners in charge. If a company has a good deal of prestige, verify that those professionals are still in the picture, and in command.

Evaluate Your Candidates

You have completed the interviewing process and have hopefully taken the time to check out their references. Now it's time to consolidate your information to make a selection. I recommend evaluating each finalist using three criteria: experience, communication ability, and teamwork. Finding a professional who is strong in all three categories will go a long way toward putting you on the right track. Others may have only one or two of these three characteristics.

Probably the most important trait to examine is experience. Firms with a longer history are a surer bet than newer start-ups. Remember that construction can develop problems as the years pass. You need a company who will stay in business and stand behind warranties or be financially responsible for errors. Obviously, older firms are in a much better position to fulfill this requirement. A good rule of thumb would be to establish five years as a minimum for a firm to have been in business. After this length of time, a company will have either folded or developed a good financial and client base.

We have previously discussed matching your project's size and scope with their typical project. If a firm has never designed or built a project such as yours, don't become its guinea pig! There is nothing worse than allowing a professional to learn while working on your house. Just as baseball players aren't ready for the big leagues without gaining experience in the minor leagues, architects and builders aren't ready to design and build major projects unless they have on-the-job experience as well as training. You want a pro, not a rookie. Determine a professional's bread-and-butter project, one that he knows backwards and forwards. This is crucial for your project's success. Stick with his strong suit! I can't think of any more cliches to get my point across!

Our second criteria is communications. During your interview, did the candidates carefully listen to your goals, requirements and ideas? Were they taking written notes about the important issues? Did they return telephone calls promptly or did you have to call them twice to get a reply? My first rule as a professional is to set a positive tone with each potential client. I return their calls promptly, and follow up diligently with the materials I promised. I appear at scheduled appointments on time, always trying to make a good first impression. If a professional puts forth a lackluster effort before he has the job, what can you expect during the project? Chances are his performance will not improve over time; in fact, it will probably worsen.

Since we have previously stressed the importance of good communications in chapter 6, you should be able to judge whether the professionals on your list will meet these criteria. On the other hand, too much of a good thing can be harmful. If you can't get a word in edgewise because the professional does all the talking, chances are he will lead you by the nose, telling you what "you must have." Communication should always be a two-way street.

Another trait you want in your building professionals is teamwork. Will they work well with others? Building projects involve a large cast of players to accomplish thousands of complex tasks. The likelihood for errors and resulting conflict can be great. Avoiding the pitfalls of adversarial relationships is very important. A good architect, builder, or any other professional encourages teamwork.

No one should appoint himself as king of the hill. Everyone has a role to play, regardless of importance, and everyone should be treated with respect and courtesy. Nothing ruins a project more quickly that arrogance. Believe me, there is enough glory on a project to go around. Ego trips aren't necessary. Evaluate professionals along these lines. Is this someone you could be friends with if the circumstances did not involve your project? This may well be the ultimate question.

Making the Final Choice

Once you have narrowed the field to a short list, request each candidate to prepare a written proposal. Review our discussion in chapter 5 to evaluate the specifics of each proposal. Make sure you thoroughly understand the exact services proposed and the associated fees. If you feel the price tag is too high, consider deleting some services to save money. Also, try negotiating the amount of the fee or construction cost. Just as in buying a car, bargaining is commonplace in the building industry. You have nothing to lose by trying. The worst answer you can get is a simple "no."

Don't forget to request a sample contract for review. Recalling my point in chapter 5, learn the exact contents of the final agreement before you commit to a prospective company. If certain elements pose a problem, discuss options or modifications with the professional. Candidates who have the qualities of our three evaluation criteria should understand your concern and be willing to find a compromise. If they are unwilling to bend over contract issues, they may well be just as inflexible about design and budget topics.

Try to equalize all proposed services and fees for each finalist. Can you confidently say that all proposals are on an "apples to apples" basis? Get the necessary clarifications and explanations, even if it means an additional interview with one or more of your candidates. Your ultimate selection has a lot to do with what type of homeowner you are, as discussed in chapter 2.

After qualifying each firm using all the facts and figures you painstakingly gathered, save your intuition for the final consideration. Can you form a trusting relationship with these people? You will be trusting them with a large amount of your money, and interacting with them on a daily basis for many months. Especially with an addition or remodeling, would you feel comfortable inviting them into your home among your family and possessions?

Your homework should give you the information needed to match each professional with your personal preference. If the fees, services and personalities are a good fit, make the call!

Chapter Seven Recap

- **Locating and interviewing the right building professionals is the first step in hiring the right team.**

- **Gather as much information on each candidate as possible, including references from friends and past clients.**

- **Interview each candidate and meet the staff that will actually be working on your project. I recommend talking with professionals from both the Traditional and Design/Build Systems.**

- **Evaluate your candidates through three criteria: experience, communication and working well with others.**

- **Visit completed past projects and talk with these homeowners to learn of their experience with each professional.**

- **Once you have narrowed the field, ask each candidate to submit a proposal for their services for your review.**

Chapter 8

Create a House Uniquely Yours

Now the fun begins! The planning and design phase you have long awaited has finally arrived. Creativity can start to flow, translating your goals and dreams into three-dimensional reality. But before we are ready to review the design your architect will produce, you must first be prepared to communicate your ideas and requirements.

Years ago as a young architect, I received my first large, custom-built house commission from a friend of a business acquaintance. For the time period during which it was built, the design was ambitious and the budget was ample—an architect's dream come true! The site was a 3½-acre wooded lot that gently sloped in all directions from a centrally located hill.

Upon meeting the couple, who were "empty-nesters" building their first house from scratch, I began to learn the dynamics of their relationship. The husband was a successful businessman who told me, "Give my wife anything she wants, as long as I get a three-car garage, a full basement with extra-high ceilings, and make sure the cost doesn't exceed my budget!" He then proceeded to leave the house design to me and his wife. When we started working on the preliminary design, his wife could not begin to tell me what she wanted. I did plan after plan based upon requirements that were constantly changing.

When the two of us started to revisit earlier plans that she had rejected, it finally dawned on me why she was so indecisive. Her husband, not wishing to become involved in the design process or just too busy, put the weight of the project entirely on his wife's shoulders. She was too scared to make any decisions that he ultimately might not like. It became my job to read his mind and insulate her from any blame for the design decisions!

Put on Your Thinking Cap

Good residential architecture is the result of productive communication between the homeowner and designer. Asking an architect to be a mind reader, as in the above story, is like telling a painter to pick any color for the walls he fancies. You will no doubt be disappointed with the finished product the painter selects. Unless you are the type who has no ideas and prefers to leave all the decisions to your architect, you must communicate your ideas. The best possible exchange of ideas is most likely to result in a new or remodeled house that meets your expectations.

This chapter takes that process one step further by helping you to decide on the interior and exterior design of your new house and its relationship to the site. This planning phase can be exciting because the project you envisioned begins to take shape. It can also be frustrating because you have so many decisions to make. Before starting to work with your design professional, get yourself organized. Much of what you will want in your new house is a reflection of your personality and lifestyle. A well-designed home should be a personal extension of you and your family. Who are you? How do you live? Answering these basic questions about yourself and your family is the first step in creating your new or remodeled house. Your answers suggest the sort of house you need, whether you're remodeling or building new.

A few examples will help to get you thinking about the special needs of your own family. Growing families need lots of room and don't mind climbing stairs, while empty-nesters may want to live on one level and prefer a smaller home. People who like birds and gardening want a home that's open to nature and surrounded by trees. Those who entertain often want a clear separation between public and private areas in their home. Family members who pursue individual interests will value individual privacy.

Asking yourself some easy questions may be the best way to begin. How long do you plan to live in this particular house? Is it a short-term stay or will it be indefinite? If you plan to live here for a long period, will the number or composition of family members change? The responses to those basic questions will immediately affect your planning. Short stays in a house may not justify the expense required to make it the perfect living environment. Longer stays may encourage increased present or future spending. For example, a change in family composition, such as children moving out or parents moving in, could require future remodeling to alter living spaces.

Encourage family members to list what they want or expect in their new home. Compile a list of rooms and amenities each family member would like. Make it an inclusive process so everyone will feel an important part of the project. Make it clear, though,

that everyone will probably have to make compromises during the planning stage. For example, explain to older children and teenagers that they'll have to choose between an extra-large bedroom or a smaller TV room, just as you have to decide between a larger living room or a bay window with seats in the dining room. Compromises are part of planning any building project. If everyone could plan their home without making compromises, many more families would choose to live in mansions!

Answering another set of questions will help you decide how you want your house to look. Will the house be formal or informal? Do you want to impress guests, or are you more concerned about making them comfortable? Do you want a house that expresses your personality, or one that will influence how others see you? For many people, these questions are a bit more difficult to answer than the first set of questions. Give them serious thought, because your answers will influence every design decision, from the exterior of your house to your choice of wall coverings.

Do all these questions make you nervous? Feel a little intimated or overwhelmed? Don't sweat the details at this stage. This is the time to think broadly and boldly. At this point in the game, it is better to loosen up than to be too structured. Believe me, you will have plenty of opportunity to tighten up and scale back later! Start with the easy assumptions and use it as a foundation to build from. Can't get past a certain issue? Skip the hard ones, and revisit them later.

Programming

The next step in planning your new house or remodeling project is communicating your decisions to the architect, architectural designer, or builder who will accomplish the design work. Architects call this process "programming," which simply means writing down what you want in your house.

You can begin your list with general comments about how you want the interior and exterior of your new or remodeled house to look. See figure 8-1 (page 90) for an example of an owner's program for a family of four with two teenagers.

Given the range of choices you have, especially at the beginning of your building project, it's normal to feel indecisive. Start with the easy, obvious choices, and work your way through to the more difficult ones. List the names of each room and the types of activities each space will accommodate. Besides the traditional living room, dining room, kitchen and family room, you may want to add more specialized spaces. For example, a big screen home entertainment theater viewing area, a computer desk, or a quiet reading space could be on your list. Also, include the amenities you want, like the hot tub for eight, a gourmet kitchen, or a bar to entertain ten friends.

Don't worry about advanced details at this stage. If you can't make up your mind where the dining room should go, skip that decision for now and describe the laundry room you know you'll need and the screened porch you've wanted for years. And don't let writer's block slow you down. You don't need a literary masterpiece, just simple statements you and your building professional can understand.

Once you get started on your list, you'll find creating your dream house is a lot of fun. But remember that every item on your list, every decision you make, will eventually translate into construction cost. Temper your enthusiasm with common sense by keeping in mind your answers to those questions about who you are and how you live. This relates to a very important entry in your program—your preliminary budget. We already discussed how you can approximate costs at this stage.

You may not know how much the project you envision would cost, but you can list the figure you can afford. I recommend working within a range. Consider an amount on the low side that would be financially comfortable. Then proceed to the upper limit that would be stretching the family budget. This

spread will start you focusing on the financial side of the project equation.

How Big Should a Room Be?

As you are considering space versus budget, consider how rooms actually function. Remember you must pay for every additional square foot! Every room fulfills three separate functions. It accommodates the necessary furniture, provides walking space to reach adjacent rooms and provides exterior wall space for windows to admit light and outside views. Identifying what types of rooms are needed—living room, bedrooms, family room—is seldom a problem. Deciding the size of each room can be a challenge. You may know you want a "big" family room or a "medium-sized" living room—but the real question is exactly how big? Keeping the following points in mind will help you decide.

While large rooms appear spacious and impressive, more floor space usually requires more furniture and more windows to balance the amount of natural light. Hence the larger the room, the more expensive it is to build and furnish. No matter how well you plan, compromising on room size is often inevitable, because of your budget or the space limitations of your lot. Keep in mind that rooms designed too tightly will appear even smaller after furniture has been placed. Fortunately, rooms that look too small for the amount of furniture, circulation, and windows can be made to look bigger by using mirrors, pale colors, and other visual tricks of the decorating trade.

An inventory of furniture and equipment you plan for each room is very useful when you begin room planning. The best way to determine the proper room size is to plan the furniture, window, and circulation functions simultaneously. When beginning to place furniture, first take into account any fixed items, such as fireplaces, doorways to adjacent rooms, and windows that take advantage of good light or a particularly nice view. Equipment that relates directly to

Figure 8-1
Owner's Program

Prepared by Mary and Joe Client

General:

- House should have an informal sophisticated image, with understated interior spaces to accommodate eclectic elements and furnishings.
- Exterior design should have an architectural statement, but is less important than the interior.
- Preliminary residence size is 3,500 sq. ft. Budget, including landscaping and professional fees, shall be between $110 - $120 per sq. ft.
- Joe will provide covenants, restrictions and survey information from his files.
- Preliminarily, two floor levels are anticipated:

Main Level:

- Entry
- Great Room, including kitchen, breakfast/work area, dining area, seating area, and entertainment center area.
- Laundry Room
- Master Suite
- Powder Room
- 3-Car Garage with possible future apartment above.
- Office/Exercise

Lower Level:

- Recreation Room – pool table, dartboard.
- 2 – 3 bedrooms, one with attached bath.

Exterior:

- Plan area for future outdoor pool and platform tennis.
- Exterior materials should be low maintenance.
- Multiple access points from house to exterior.

General Interior Items:

- Tall windows and tall ceilings.
- Level changes are acceptable – provide wide stairs to accommodate future "stair-walker".
- Consider accommodation for "live-in".

Great Room:

- Kitchen should be mostly open to Great Room. Laminated cabinets with Corian™ or stone countertops shall be used. Vented appliances shall be used.
- A Breakfast Room area adjacent to the Kitchen shall also be utilized by Mary as a work space. Built-in shelving shall accommodate books.
- Built-ins shall accommodate glassware and other serving pieces.
- Dining Area shall seat 6-8. Additional expansion shall overflow into Great Room.
- Fireplace can be either prefabricated or masonry.
- Entertainment Center shall accommodate big-screen TV, stereo, VCR, etc. with doors that will close-off opening. Access to the back would be useful.
- Wet Bar with undercounter refrigerator.

Master Suite:

- Bedroom shall have minimal furniture.
- Dressing Area with built-ins will take the place of traditional dressers. A dressing table will also be designed.
- A large common walk-in closet with a center island will be located off the Dressing Room.
- The Bathroom will be amply sized, to contain his/her sinks, toilet compartment, shower compartment with bench and two shower heads, and a whirlpool soaking tub.

Joe's Office:

- Located off the Great Room with built-in work surfaces, shelving, files, ample counter space.
- Exercise Room, approximately 8' x 10' located off the office.

furniture, such as televisions, VCRs, and sound system equipment will also influence where you place furniture groupings.

Some rooms have special requirements. Here are a few rules of thumb. Beds and nightstands are usually located on a solid wall with no windows. Living rooms and family rooms should be sized according to how many will commonly use the room. Place furniture so that each person can see and talk to everyone in the room. Dining rooms are sized according to the number of people you wish to seat. Measure the dining room table and add four feet to each side of the table for chairs and circulation, plus extra space for furniture pieces placed against the walls.

Several helpful guides are available if you want to try hands-on room planning. "Cut-out" furniture kits, with pieces sized so a ¼-inch equals one foot, can help you begin to visualize room design. For those who are comfortable using a home computer, software packages such as *3D Studio* and others are also available to plan room sizes.

From Programming to Preliminary Design

Your program explaining what features you want in your new or remodeled house and how you and your family will live gives your designer the basis for a preliminary plan. It's also used for estimating the total size of the house and arriving at a construction budget.

By knowing how much you want to spend and the size of the house you want to build, the designer can arrive at an estimated budget. Construction costs can be calculated using the guidelines in chapter 3. Before preliminary planning begins, ask your designer to help you estimate the cost of the project you want to build. This will help you to keep your plans within your budget. For example, you won't expect to build a new house with 4,000 square feet of space if your construction budget is $200,000. This would allow only $50 per square foot for construction—too low

a figure in today's market. A more realistic figure for an average house is $100–$125 per square foot.

The more your program list tells your building professional about what you want in your new or remodeled house, the more readily he or she can come up with a first design that reflects your goals and budget. If you have balanced your space requirements and available budget, your designer shouldn't present you with a plan that is twice the space you need and double the cost.

How does the architect or architectural designer use your program list? By applying problem-solving skills and the ability to think in three dimensions, he works to create a house plan that contains the rooms you and your family envisioned. He or she may also suggest different ideas based on knowledge and experience gained from similar projects.

To help you create your new or remodeled house, your design professional considers both form—the way the house will look, and function—how the rooms and their arrangement will meet your family's everyday living requirements. The rooms should be logically interconnected and proportioned to the total area of the house. The plan should also harmonize with the building site.

Site Planning

This is a good place to introduce the importance of the site (better known as your lot) in preliminary planning. Often homeowners are so focused on the physical aspects of the house, they forget that it is sitting on a piece of land. Whether your site is big or small, it should become part of the design.

Before you go any further with planning, you need to understand the importance of the site in creating a new house or one with additional space. (See chapter 13 for site considerations if you're remodeling or adding on to your house.) The house plan should evolve from the characteristics of the site. The best architecture has always been created by letting the site features shape the design of the house.

According to legend, Frank Lloyd Wright often brought a chair to an empty building site and sat for hours to get a "sense of the land." On larger, more rural sites he used to camp out for days. Unfortunately, today's residential design often forces a house plan to conform to any given site, regardless of how poorly it may work. Sensitivity to the land has unfortunately taken a back seat to building larger and more opulent houses on smaller lots.

Trying to force a favorite house plan on a lot often leads to redesigning parts of the house. This in turn leads to compromises and poor design. You can avoid these problems by choosing a building lot first, so that the house can be planned to suit the characteristics of the site. If you feel you must have a specific house plan, be prepared to search for a site that will accommodate that particular design.

For an example of the cart before the horse, let me tell you about my visit to a newly completed project. As I drove up the driveway, it seemed that I was looking at the back of the house instead of the front. Upon asking a few innocent questions, I discovered that the owners were so enamored by a floor plan of a house they had seen, they turned the house around to fit the site orientation instead of redesigning! The house had no curb appeal whatsoever, but the owners thought their house was wonderful. Wait until they try to sell it!

Every building site should be analyzed for a number of parameters before the architect begins his planning. You can do a simple analysis of a potential lot by standing in the center and slowly turning a full circle. Perception is always in the eye of the beholder. Do you like what you see, or are there too many negatives? Many building sites have positives and negatives in different directions. From one direction you may view adjacent houses that seem compatible with the one you plan to build. Looking in another direction, you may see a busy four-lane highway with lots of noisy traffic. Should you eliminate this lot from

your list of potential sites? The ultimate decision is yours. Use your built-in radar to carefully consider every positive and negative aspect of each possible building site. Don't trust your memory here. Make a detailed list of your impressions of every lot you visit.

Several criteria should also be considered when evaluating vacant land. The grading or topography (slope) of the site can have a major effect on the feasibility of a basement or level changes in the floors. Sun patterns exposing the house to morning and afternoon sun can influence the location of principal rooms. Wind patterns affected by trees or other natural screens may determine window and door placement to enhance cross ventilation. Views both looking *at* and *out* from the site are very important. Take advantage of positive elements, such as a beautiful stand of trees, or minimize the negative elements, such as power lines or unattractive views of adjacent houses.

The very shape of a lot can dictate the layout of the house. Not all lots are shaped in nice, rectangular proportions. Often an irregularly shaped site can force the designer away from conventional solutions and into new creative forms. Simply considering the best location for a driveway entrance in relation to other properties can turn a house plan completely around.

Site Problems You Can Avoid

A site may need more research if you will be one of the first to build in a given area. A visit with a staff member at the local municipal or county office can provide very useful information about future use of the land surrounding your potential building site. Will only residential development be allowed in this area, or is a shopping center planned a few blocks away? Does the county plan to expand the four-lane freeway a quarter of a mile away? It's surprising how many homeowners buy lots or houses without bothering to ask these basic questions.

For example, a client for whom my firm was designing an office building complained about the house he had just purchased. Instead of investigating the perimeter area of the large lot himself, he took the real estate agent's word that he was buying a home in a peaceful, quiet neighborhood. The day his family moved in, they discovered train tracks close to the rear of the property! I know this sounds hard to believe, but it is a true story. The last time I spoke with him, he was looking for a lawyer, hoping to lay the blame on someone other than himself.

Even small, flat, open subdivision lots require a closer look to avoid possible problems. These lots appear to be almost identical to one another, but if you check carefully, you'll find important differences. Consider traffic patterns in the development. Lots close to a main road entrance may be noisy and dangerous for children. Corner lots have these same problems and may not have a well-defined back yard. Look for a lot on a cul-de-sac or a street with no through traffic, away from the entrance to the subdivision. There are also important differences in the shape and dimensions of different lots. Some have narrow fronts with longer depths. Some are more square, or have peculiar angles. Others could be surrounded by four or five adjacent lots. Try to get a lot that is more rectangular than square, and without property lines that come to a narrow point. Consider also how close neighboring houses will be.

It's also wise to avoid lots that are on the perimeter of the subdivision, unless the adjacent land has been fully developed to the extent that it won't change. Vacant perimeter land could put you adjacent to future commercial real estate developments or other less desirable residential developments.

Choosing an Exterior Style

Once you've made a good start planning the interior of your new or remodeled house, you can think about how you want it to look from the outside. No rule says you shouldn't consider the exterior first, but most people work from the inside out because they feel uncertain about choosing an architectural style.

There's good reason for uncertainty. Few newly constructed houses follow any true, pure architectural style, so it's almost impossible to give a new home a true stylistic label. Most of today's designs can be called *neoeclectic* because they revive or adapt the best elements from past architectural styles. Current traditional house designs are usually free adaptations of older, purer styles that originated between 1850 and 1940.

The wide variety of roof forms, windows, and ornamental trim appearing on current houses makes it difficult to pin down a style. Confusion abounds, especially when builders and real estate agents use simplistic terms to identify what they sell. Frequently, traditional houses are called colonial for lack of a better term. In some respects, many elements of these houses have roots in colonial architecture, but it's more accurate to say that they are derivations of many other styles. Take a seat and attend my short course on residential architectural history.

A brief look at American house architecture shows four distinct periods. The earliest phase is the original colonial era, around 1600–1820, with English, French, Dutch and Spanish regional influences. The second period, between 1820–1900, has examples of classic-style revivals, such as Greek, Gothic, and Italian, as well as Victorian styles. The third phase, from 1900–1940, is an eclectic era, where all of the preceding styles were copied, but usually simplified and more informal, sometimes borrowing from related styles. Finally our current period, neoeclectic, borrows from everything available, producing a mixture of less pure styles.

To get some idea of the style of a house, start with the overall shape of the floor plan, form of the roof, and then work toward individual elements. For example, gable roofs are English colonial, barn-shaped roofs are Dutch colonial, front porches are French colonial, and projecting bay windows are Gothic Revival. Turret roofs and arched windows are Victorian Gothic, and round or square columns are Greek Revival. Low-pitched roofs and small-paned windows are Georgian.

Our four periods in American architectural history encompass more than forty different design styles. We have a lot to borrow from, making it difficult to classify current house designs. You probably already have some ideas about how you want the exterior of your house to look. A drive through residential areas may help you expand on these ideas. Even choosing elements of certain houses in combination can be a good start. Turning to the broader context of the entire house, you can also look through architectural or home magazines for pictures of design elements you like. By clipping magazine pictures, you can accumulate a file ranging from entire houses or rooms to specific items, such as a doorknob or a fireplace. This will give the person designing or remodeling your house a clearer understanding of your tastes. The statement "a picture is worth a thousand words" certainly applies here.

You can find pictures of rooms in a number of magazines, including *Home, House & Garden, This Old House,* and *Architectural Digest.* If you can access the Internet, you may find information and pictures of plumbing fixtures, cabinetry, appliances, lighting fixtures and other "finishes" on manufacturers' Web sites.

If you want a "purer" architectural style, *A Field Guide to American Houses* by Virginia and Lee McAlester (Knopf, 1984) and *Identifying American Architecture: A Pictorial Guide to Styles and Terms, 1600–1995,* by John G. Blumenson (AltaMira, 1995) are excellent sources. Even if you choose not to follow a pure style for your project, you will have learned something useful for your next cocktail party!

Pros and Cons of Stock Plans

Homeowners-to-be who are uncertain which style of house or floor plan to use often turn to myriad "stock plan" books available at bookstores. Hundreds

of plans with different floor configurations and exterior styles are shown, and for an additional fee you can order architectural plans for any model. If you find a house style and plan that fulfills your requirements, you have taken a big step in saving money, as these plans are far cheaper than hiring an architect.

While these stock plan books may be a good first step in broadening your knowledge, beware of certain shortcomings if you choose to purchase plans to use for actual construction. The designers, who often do a good job with the overall house concept, can put you at a disadvantage. Since these plans are off-the-shelf and intended for sales across the country, they don't incorporate local conditions that may affect the design. For example, the designers have no knowledge of your local building and zoning code requirements. Setbacks, the minimum distance that the house must be placed from the front, rear and side lot lines, will affect the shape and size of the home you may legally build on your lot. Consequently, a site plan showing the house placement will have to be prepared locally for the building permit. Also, remember that the site where your home will be built has particular characteristics that may affect the room layout. For example, solar orientation—which direction the windows will receive light and lose energy—determines window placement. You will also want windows to take advantage of landscaping and open space.

Topography, or the contour of the land, may require either a walk-out basement or placing the driveway on the opposite side shown on the stock plan. Native trees or vegetation worth saving can also affect the placement or floor arrangement of certain rooms. Another possible problem is that stock plans often don't provide a foundation plan, because requirements vary widely from one region to another depending on frost and soil conditions. The depth of foundations and their structural design can vary greatly from one locale to another. Here again, you'll need local expert help.

All metropolitan areas traditionally have building codes, especially for electrical and plumbing work, that are more restrictive than the rest of the country. If you expect to use stock plans for construction in one of these more restrictive areas, be prepared to add thousands of dollars more in planning and construction costs to your building budget.

Another problem that can crop up with stock plans is that certain areas of the United States, particularly large cities and suburbs, require architectural plans to be sealed by an instate architect. Plans prepared by an out-of-state architect may not be accepted. In many states it is illegal for one architect to extend his license for work that was prepared by another architect. Architects can lose their license if they are caught authorizing another architect's drawings without permission. Another concern is that stock plans are usually copyrighted. Removing the original architect's name and substituting another architect's name would violate this copyright.

If you prefer to use a stock plan you like very much with no changes, locate a builder or architect and general contractor who can furnish the missing information required to obtain a building permit.

If room layouts and sizes shown on the plans you choose aren't exactly what you need, changes may be necessary. Something as seemingly simple as moving a wall on a stock plan could cause changes in the structural, plumbing and heating work. I've found that if you are going to change more than 20 percent of a stock house plan, it's usually easier for both the owner and builder to start from scratch.

Despite these drawbacks of stock plans, you may still find them useful. Many homeowners choose to incorporate some stock plan design elements into a locally prepared set of new plans. This hybrid approach allows you to combine some of the economy of a stock plan with a customized design that suits your building site and includes the special features you want in your house.

Preliminary Design

By the time you finish your program list of needs and wishes for your new or house remodeling, you may have included design features from three sources: your own ideas, pictures and stock plans, and your architect or builder. To help you visualize the home you're planning, whoever is designing your house will first present a preliminary design. This may include a site plan, floor plans, or even a model, that shows the house form as it sits on the site. See figure 8-2 through figure 8-4 for our example of a new house. Notice that the house in this site plan has been positioned in a clearing near the slope descending to the back of the lot. The floor plans show the basic layout of rooms, their sizes, windows, doors, kitchen, bath and entertainment features.

Don't be surprised if your first look at preliminary plans for your house is both exciting and disappointing. Exciting because you know your house is beginning to take shape. Disappointing because you may not be sure what you're looking at. How is anyone supposed to visualize the design from a series of confusing lines on a drawing? You will probably find that trying to translate those lines and symbols into rooms, windows, hallways, and stairways is frustrating, because they're drawn in a specialized language.

And yet understanding the content of these plans is crucial to the success of your building project, because they convey all of the design ideas for your house. Some homeowners make matters worse because they're embarrassed or too proud to admit to their design professional that they have no idea how to read the plans.

At an American Institute of Architects convention, a successful architect told me he almost gave up after designing his first house. "My clients seemed to like everything, from the first drawings to the final plans. They never questioned anything—they always told me what a good job I was doing. But when we walked through the finished house, they acted as if they didn't recognize the design.

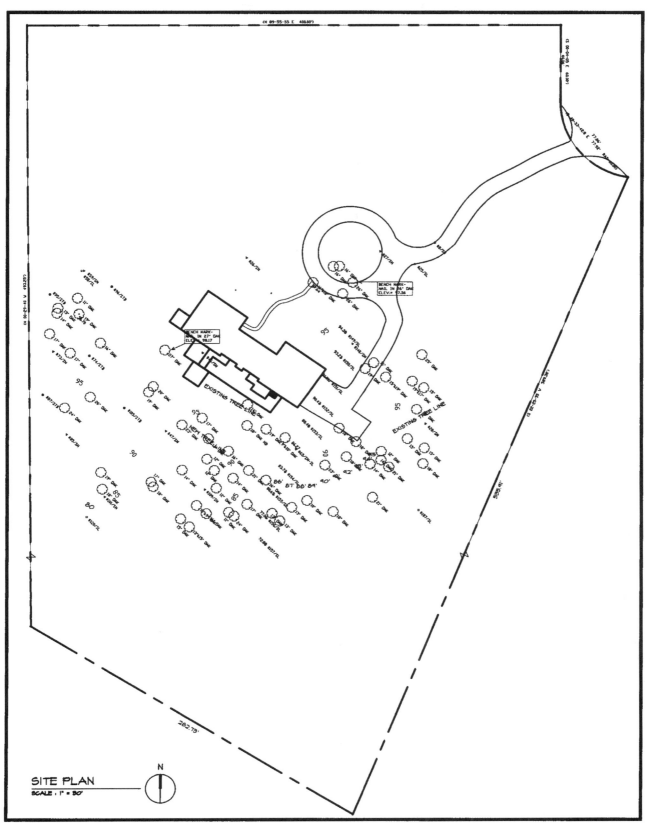

Figure 8-2
Preliminary Site Plan

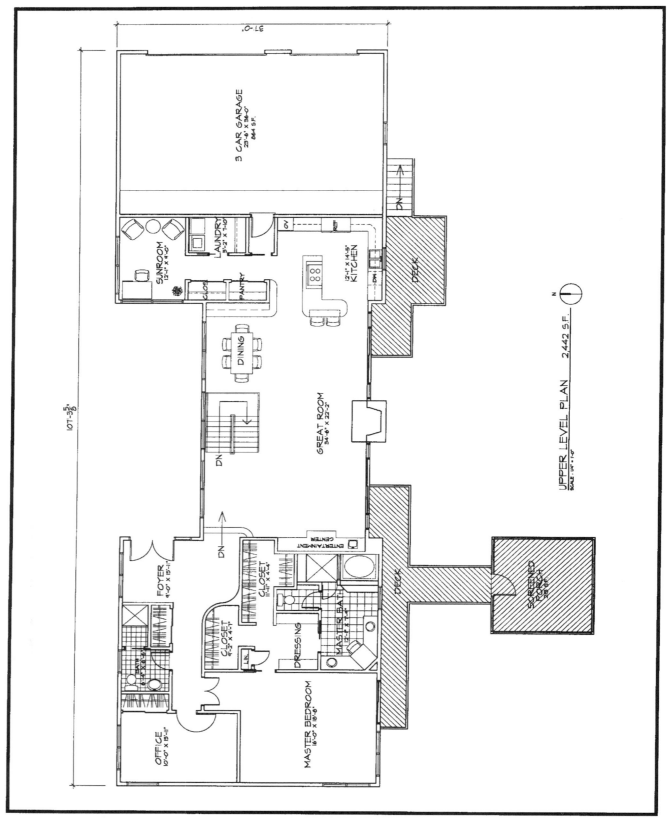

Figure 8-3
Preliminary Floor Plan

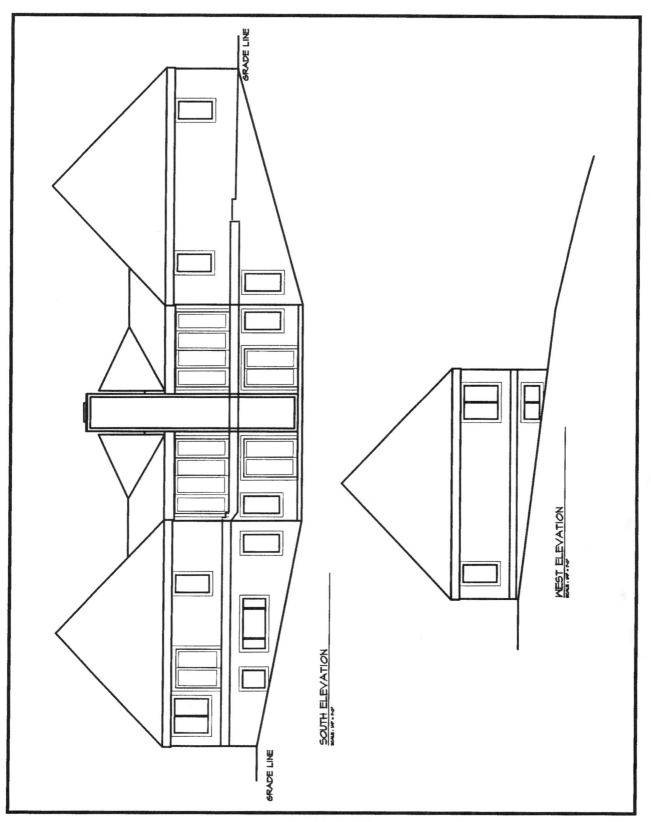

GRADE LINE

WEST ELEVATION
SCALE : 1/8" = 1'-0"

SOUTH ELEVATION
SCALE : 1/8" = 1'-0"

GRADE LINE

Figure 8-4
Preliminary Elevations

"Before we got past the front hall, my client said, 'This can't be our house. It's nothing like what we wanted!' and burst into tears. When I explained it was just what the plans showed, her husband said, 'I guess we didn't understand those plans very well. We should have asked more questions.'

"That taught me I had to work hard at getting clients to ask questions. I tell them no question is too small, and I suggest they write every question down, no matter how unimportant it seems," the architect told me.

Unlike this confused couple, you must speak up and ask questions. If you don't understand architectural drawings, you have lots of company, but your financial stake is too high not to thoroughly understand the design from the beginning! Don't sit quietly and nod your approval if you have no idea what your architect or designer is talking about when he presents plans for your new or remodeled house. Even in the earliest planning stage, before your building professional has begun work on preliminary drawings, asking questions about the overall design and the cost is vital to the success of your building project. Speak up! No question is too silly or too small. You have a lot of money riding on this bet!

Help from Models, Sketches and 3-D Computer Drawings

Architects have tools that can help you get past the difficulty of understanding two-dimensional drawings. Scale models can help you to understand what is shown on the drawings. If you have a hard time trying to picture the exterior appearance of your project, your architect can build a mass model that simply shows the exterior form. A more elaborate model can show how the house relates to the site, as well as indicating the scale and relationship of interior rooms. You can instantly see your project from any viewpoint. Figure 8-5 is a *mass model* that shows the exterior as a monochromatic "solid mass." Figure 8-6, is a *presentation model* showing doors, windows and color. Three-dimensional sketches can also give you an idea of the exterior or of interior spaces that have special elements, such as high ceilings, large window areas, or special equipment.

Today's good news for homeowners is that most architectural offices now use some form of computer-aided design and drafting (CADD). New programs specializing in three-dimensional modeling are changing the way architects can help homeowners visualize their new house in the planning stages.

Instead of confining the architect's original design to two-dimensional plans or sketches, the computer can present the exterior and interior in three dimensions. Sitting in front of a computer screen, you can tour the exterior of the house from any angle you choose. Even selected interior room views can be shown. Many computer programs enhance the image of the house with the addition of shading produced by the sun at different times of the day and year.

Design elements can also be changed during your tour. Placement and sizes of windows and doors can be revised as you watch the screen. Roof shapes, such as gables and hips, can also be varied, providing different design alternatives. These three-dimensional images are not confined to the computer screen. Advanced plotters can quickly give the homeowners selected drawings of what they have viewed to take with them for reference. For an example of CADD generated perspectives, see figure 8-7. As you view the plans on the screen, building materials can be changed to show different looks with brick, stone, siding, and architectural trims. Sites with slopes and unusual topography can also be illustrated with the proposed design situated within the contour of the land.

Advances in computer visualization have also reached the retailing level. Kitchen and bath designers and distributors can take your plan, and within an hour, have your kitchen with the cabinets, appli-

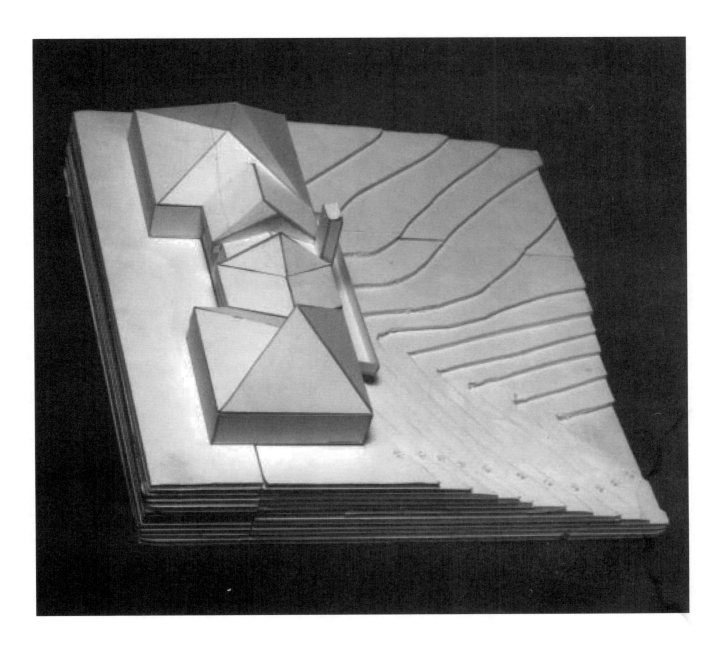

Figure 8-5 *is called a* mass model *because it shows the exterior as a monochromatic, solid mass.*

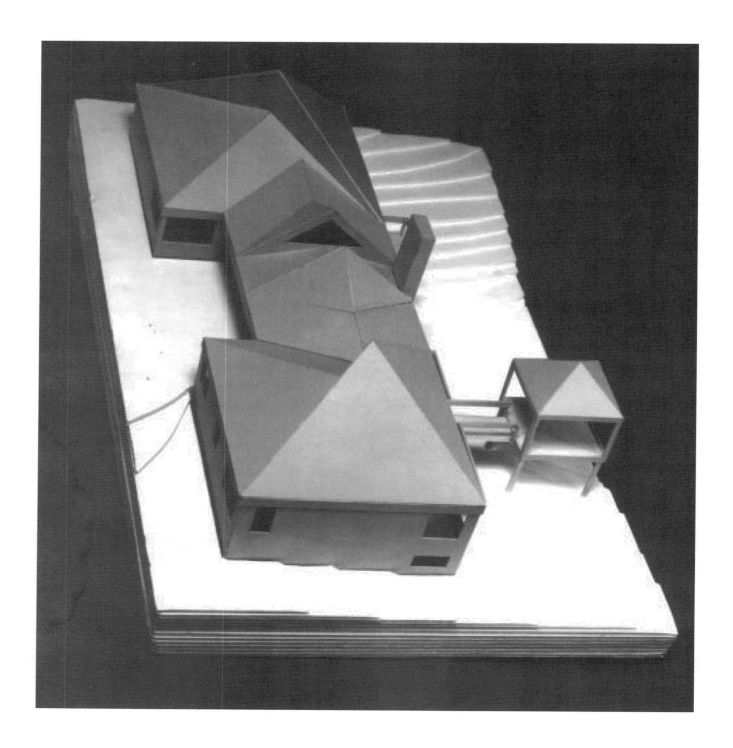

Figure 8-6 *shows a* presentation model, *which further indicates doors, windows and color.*

Figure 8-7. CADD Interior

Figure 8-8   *Interior perspective*

ances and accessories shown in three dimensions. Styles can be easily changed, and a detailed cost estimate quickly produced.

Figure 8-8 shows a hand-drawn sketch of an interior space. Many architects and designers still use this effective method of conveying design ideas and it is perfectly acceptable.

In chapter 9, which deals with design development, you will find out how a preliminary design becomes a finished plan for your project—and how you can play a role in controlling the cost.

Chapter Eight Recap

- **Good residential architecture is the result of effective communications between the homeowner and professional. Before starting work with your professionals, get yourself organized.**

- **Analyze how you wish to live with your family members and identify your goals and requirements for the project.**

- **The most effective way of communicating your project to your professionals is by a written record, called a *program*. Every issue you can address should be informally listed.**

- **Express your available project budget in a range from low to high. This will help reveal if your initial project's scope matches your budget.**

- **At the beginning of architectural design, analyze your land site for advantages and disadvantages.**

- **Architectural styles should be considered when establishing the project's parameters.**

- **Stock architectural plans can be useful for both preliminary design input and actual project use, as long as certain local building code and site issues can be satisfied.**

- **Architect scale models, three-dimensional hand drawings and computer-generated views of the exterior and interior can be helpful in understanding what your designer is proposing.**

Chapter

Design Development: Plan Ahead to Win the Budget Battle

A local builder I knew developed a reputation for great custom designs, good workmanship, and cost overruns. His recipe was always the same: Get the homeowner to sign on the dotted line by promising only the house size and price—without supplying any specifics. He would lead prospective buyers through a completed custom house, dazzling them with expensive materials, finishes and interesting architectural design. When asked how much such a house would cost, he would reply, "Well, this one starts at this amount and goes up, but I know ways of building it for less. We can work out the fine details later."

The builder would produce a custom contract that stated the number and sizes of the rooms to be built, a long list of "building standard materials," each with a certain price limit and a "preliminary" total cost. Unsuspecting homeowners saw that the rooms matched the existing house they visited, and the price was attractive, so they usually signed without a second thought.

After construction was under way, the builder would call the homeowners to his office to "select colors." He loved these meetings. By the time the homeowners left, his profit had increased by 20 percent. In his office the couple was shown samples of

cheap carpet, paper-thin doors, and low-grade cabinets. The couple quickly realized the "building standard materials" term agreed to in the contract provided far less than expected. When they questioned the builder, he answered, "I told you we could bring in your project for less. These are the materials we will have to use. If you want to upgrade some of your selections, I will be happy to show you nicer materials." He would then pull out the good stuff, eventually increasing the cost with little additional effort on his part.

At this stage of a project, *New House/More House* has taken you from receiving the kickoff at the beginning of the football game to moving the ball up the field all the way to the fifty-yard line. There is still much more work ahead to get the ball into the end zone! It has been my experience that a project's success is directly proportionate to the amount of work accomplished during this phase. Design development, if you remember from chapter 3's examination of architectural services, takes the preliminary design and adds all the details. The term "finishes" has been used several times in previous chapters. Since these items often cause cost overruns, I've devoted this chapter to helping you avoid these budget busters.

Look around the room in which you are sitting. Every surface has a finish. The floor may be carpeting, wood, or ceramic tile, the walls could have paneling, wallpaper or paint. Lighting enables you to see in the dark, doorknobs permit entry into rooms, and faucets provide running water. You may take these things for granted, but someone took the time to consider and select each of these finishes.

Choose Interior Finishes Early

I know many homeowners don't care about this level of detail; after all a light is a light and all sinks have running water. But these details account for approximately 40 percent of a project's total cost. The previous chapter discussed how much of your building dollar goes into building the first 60 percent. It has been my experience that budget busting results not from overruns in the house's shell, but from overruns in finishing the interior. This phase of the project process concentrates on perfecting the overall design and selecting the interior finishes while still keeping the budget in balance. Here you will select cabinetry, appliances, floor coverings, bath fixtures, and ceramic tile, among other things.

The length of this procedure depends on the level of your custom design. If your budget is limited or your tastes tend toward the ordinary, this step will involve merely picking colors. Your professionals will have already determined the cost of your finishes. However, if you are lucky enough to have more money at your disposal and wish to explore different custom design possibilities, design development will take much longer.

To prove my point about the importance of this process, consider the price range of certain material finishes. Carpeting can start at $10 per yard and easily reach $60–$70 per yard for top of the line. A kitchen cabinetry package can be had for as little as $5,000, or it can add up to $100,000. The stairs could be built on-site inexpensively by a carpenter with lumber and home center railings, or a custom stair fabricator could furnish the complete assembly in walnut for a dramatically higher price. Thus, your project cost can escalate quickly or be closely controlled, depending on your selections.

To keep the budget in balance, you must spend the time to complete this package before construction begins. Incomplete selections at this stage will usually result in cost extras and time delays. Another critical point has been reached in determining the fate of your project. If your professionals are not guiding you to make these choices, force them to focus their attention before proceeding further! Do not let them string out this step during construction, unless your selections are very basic. Take the time now to get it right.

Why should you do this early on? Here's an answer that all taxpayers will understand. In their infinite wisdom, Congress and state legislatures first decide how much money to spend for the year running the government. After passing many bills and throwing money in all directions, they tally up their expenditures at year's end only to find they overshot their original budget limit. We are going to run things differently, since we don't have the taxpayer to help bail us out of our mistakes! We will determine the cost and relative merit of each item before we commit to a final number. This cart will definitely not be before this horse!

Work with your builder or architect to make sure you know the quality and cost of every finish. Figure 9-1 provides a checklist itemizing these selection categories for your reference. Usually, a design professional follows a predetermined path in selecting certain major finishes and colors. This establishes a central concept that assists in choosing the minor items. I like to start with floor coverings and cabinetry before considering paint and plumbing fixture colors. By the time you have made all these selections, your project budget should become much more of a real figure.

Richard Preves & Associates, P.C.

Architecture   Planning

Figure 9-1

Design Development Checklist for Finishes

Client Residence
December 4, 2000

A.   Interior Finishes

    1.   <u>Bathrooms</u>
         a.   Plumbing Fixtures
         b.   Floor Finish
         c.   Wall Finish
         d.   Vanity Top Material
         e.   Vanity Cabinet Design
         f.   Mirrors
         g.   Toilet Accessories
         h.   Lighting

    2.   <u>Bedrooms and Workroom</u>
         a.   Carpeting
         b.   Wall Finish
         c.   Lighting

    3.   <u>Entry and Hallways, Closets</u>
         a.   Floor Finish
         b.   Wall Finish
         c.   Lighting

    4.   <u>Stairs</u>
         a.   Floor Finish
         b.   Railing Detailing
         c.   Finish

    5.   <u>Living and Dining Room</u>
         a.   Floor Finish
         b.   Hall Finish
         c.   Lighting
         d.   Fireplace Finish

    6.   <u>General</u>
         a.   Doors
         b.   Door Hardware
         c.   Window Operation, Glass Type
         d.   Built-in Cabinetry
         e.   Painting in Existing Rooms

B.   Exterior
    1.   <u>Wall Materials</u>
    2.   <u>Entry Walkway and Court Finishes</u>
    3.   <u>Landscaping</u>
    4.   <u>Front Door Design</u>

You will often be selecting these finishes from samples and catalogs, or even from less desirable photocopies of catalogs. Let me give an example I am sure you will understand. Nearly everyone has bought merchandise through a mail-order catalog. When you do, you take a chance that the item from the catalog will match your expectations when it arrives. Since there is a money-back guarantee, the worst that can happen when the item is half the size you envisioned is that you have to send it back for a refund. Unfortunately, in the building industry, money-back guarantees don't exist! Once an item is sitting on the site, you own it—unless you want to pay a 20 percent to 30 percent restocking charge. For those who major in indecision, this can quickly add up to big bucks!

To avoid this type of surprise, I recommend you spend the time seeing the options in person, especially for the major finishes. Your building team can provide actual samples or you can visit manufacturers' showrooms, home centers or completed projects with similar selections. While you are reviewing these materials, always try to get brochures or samples to keep on hand for further reference. Since choosing finishes is a cumulative process, having samples available is a big plus if you later decide to change materials or colors. For instance, if halfway through finalizing your choices you decide the current color scheme is a little off, the samples will enable you to determine which are still acceptable and which have to be revised.

One important point to remember when selecting different finishes: every material has a term we use in the business called *lead time*. This is the amount of time required to order the item and have it shipped to the site. Many materials are not always stocked, sitting on a shelf ready for your use. Manufacturers often have limited production runs once or twice a year for materials such as carpeting, ceramic tile, and brick, to name only a few. If the timing of your project does not happen to coincide with their sched-

ules, some of your choices may not be available for months, prohibiting their use.

When viewing potential finishes, always ask about the availability of each item. If you are told eight to ten weeks, add a few extra weeks for insurance. Building material delivery reliability is always suspect; available supplies can sometimes be delivered elsewhere. With the assistance of your team, screen out selections that will not fit into your project schedule. This technique will go a long way toward keeping your schedule on track.

You will soon accumulate a pile of materials, reference booklets, and swatches. Organize them into a handy file, keeping the final selections and disregarding the rejects. During construction, if a certain installed finish differs from what you expected, pull out the catalog or sample and compare. Substitutions by the architect, builder or contractor should not be permitted unless they obtained your prior approval.

Interior Designers

Faced with many finish choices and colors, overwhelmed homeowners occasionally enlist the help of an interior designer. Designers can add valuable input into the overall design. Often architects and builders do not spend adequate time considering how furniture and living patterns will function within the house. The finer details of closet design such as rod and shelf space or where clothes ready for the laundry will be stored are rarely anticipated. Habits such as shaving and applying make-up vary among homeowners. Your living environment should serve your daily routine. Without careful consideration, you could be forced to adapt to a home not designed to fulfill your everyday needs. Beyond selecting finishes and colors, interior designers can assist with these issues that you may easily overlook.

You should hire an interior designer as carefully as the other members of the project team. We mentioned a "triangle of conflict" in describing the Traditional System, but a "quadrangle of conflict" awaits you if you hire an interior designer who doesn't work well with others. Match the interior designer with the members of your team. Remember the qualities we have discussed. The designer should be a good listener and communicator and have the requisite experience. Don't expect to hire a designer and have the architect or builder direct his or her work. You will be partly responsible for coordinating the efforts of all your design professionals.

Two concerns await you in integrating the services of an interior designer into your project. First, you must clearly understand his specific design responsibilities. Since the designer's role has greatly expanded from the traditional "decorator" label of the past, a clear line of demarcation for design responsibility must be established. In addition to selecting materials, colors, and furniture, designers frequently will help shape the size of rooms, as well as kitchen design, lighting and bath design. Homeowners frequently become confused as to which services the architect and interior designer will provide. Since these services can often overlap, confusion is waiting right around the corner.

Your goal is to have the architect and interior designer thoroughly understand who does what. The last thing you want to hear on your project is, "It's not my job!" This means certain design duties fell through the cracks, with neither designer being responsible. On the other hand, it is foolish and expensive to pay both to design the same part of the project. How do you solve this problem? One method I recommend is placing the responsibility for all "permanent" items with the architect or builder. This would include cabinetry, lighting, plumbing fixtures, flooring, and hardware—items that do not change. The designer in turn would be responsible for all "nonpermanent" items such as window and wall coverings and furniture. This arrangement gives most of the responsibility to the architect or builder. Another scenario could have the architect accountable only for the general design and structure while the

interior designer would specify all the interior finishes.

During interviews with the architect and the designer, you should find out which division of responsibility each of them prefers. Regardless of the final arrangement, it is your job to carefully pair up the design team so that services and personalities properly mesh to produce a successful project. You can accomplish this task by a careful review and discussion of each design professional's proposal.

Timing is the second concern you must address. Unfortunately, homeowners who hire an interior designer often do so too late in the process. After thinking they can make all the material and finish selections on their own, or trying to save money by avoiding additional design fees, homeowners finally admit their inadequacies and hire designers right about the time construction begins. My experience on a past project illustrates this problem.

I served as an architect for a new custom home with an ample budget. My clients and I had progressed through the design and design development stages, easily selecting the colors and finishes. Construction was now moving along nicely with the wall framing almost complete. My clients, who had just returned from a short vacation, called me early Monday morning expressing concern with their interior selections. It turns out they had stayed with friends who convinced them they must have an interior designer to do justice to their new home. Although I tried to convince them it would only create problems at this stage, they insisted. As they interviewed prospective designer candidates, construction progressed, using the original design.

Two weeks later I was requested to attend a conference with the designer they finally selected. Meeting at the site, the designer proceeded to demand changes to the plans, requiring relocation of walls, windows and doors. Instead of reusing their present furniture as planned, the designer had convinced my clients to buy new furniture, which changed how the rooms would be used. I was not upset by my clients' wish to make changes, but I reminded them it would be expensive to move walls and openings already in place. The designer threw up his hands and said the cost would have to absorbed. He could not live without the changes. My clients were stuck in the middle, trying to decide whether they should spend the extra money and accept a construction delay.

Interior designers should be involved early in the design process to avoid these problems. That means you will have to predict your need for interior design help at the beginning of the project. Remember, it is far cheaper and faster to move a wall on a drawing than once it has been built. Homeowners who are decisive and know their own minds may not need this additional assistance. Those who are indecisive or cannot agree with their partner are good candidates for early interior design help.

Beware of Cost Allowances

This is a good time to introduce a topic that often causes many problems: *cost allowances*. This term simply means that instead of selecting the actual finish now, a dollar amount is reserved to purchase a selection in the future. Considering the large number of design decisions a homeowner must make, cost allowances can be a good concept that moves the project along when time is getting short. If applied correctly, this idea is a useful aid for the homeowner. But financial disaster is nearly guaranteed when cost allowances are misused.

Cost allowances can appear at frequent points along the way. An architect, trying to complete plans for bidding, may turn to allowances when the homeowners have not made all their selections. For example, suppose I have a deadline to submit plans for contractor bidding on a project, but the kitchen appliances have not been selected. Within the drawings, I instruct the contractors to include a $7,500 "allowance" for appliances within their bid. When

my clients finally make their selections, they have a $7,500 credit available. If they spend more, the difference comes out of their pocket. Spending less rewards them with a refund from the contractor. Thus, I am able to keep the project on schedule, as long as my allowance estimates meet my clients' expectations.

General contractors also routinely use allowances. If information is missing from the drawings, generals will plug in an allowance to complete their bid. Or, as a bid deadline approaches and the contractor could not get the cost of the specific hot pink whirlpool tub from the manufacturer in time, an allowance estimating the cost is reflected on their proposal. Again, this technique keeps the project moving, but could create a problem if the allowance amount is too low.

Builders are especially fond of cost allowances because they enable the builder to put together a contract for your signature before you have an opportunity to go through the material selection process. Using this procedure will benefit the homeowner only if the builder has integrity. If allowances are predetermined at an acceptable level, the project should be in good shape. On the other hand, a budget debacle awaits you if the contract includes allowances that are well below your expectations.

The following is a typical project scenario, much like the story in the beginning of the chapter. The builder's cost allowance for the entire cabinetry package in a new house is set at $10,000. For the uninformed homeowner, this amount sounds like a comfortable number. The first surprise occurs when the homeowner selects bathroom cabinets totaling $2,500. This means there isn't enough funds left to furnish the cabinetry for the kitchen. Any difference above the $10,000 allowance is an extra cost coming out of the owner's pocket. Some builders deliberately use low cost allowances to make the initial price of the project more attractive. This tactic is very similar to spotting great prices on new cars in the Sunday newspaper, only to find a stripped down model when you visit the dealer.

Unfortunately, the cost allowance nightmare does not end at this low point. Since the builder has your signature on the contract, you are captive to his pricing procedures. This additional maneuver enables the builder to charge whatever price he wishes for upgrades you select above the base allowance. You could be charged one and one half to two times the fair market value for material upgrades. When comparing builder proposals, beware of the too good to be true price. An unscrupulous builder may be anticipating built-in allowance overruns to inflate his profit as the project proceeds.

Accurate cost allowance amounts should be established at a level commensurate with your tastes, expectations, and budget. Cost allowances may be used for almost any finish material. They come in many different sizes and shapes; expressed as a total cost, a cost per square foot, or as a unit price. Common allowances include floor coverings like carpeting and wood, cabinetry and counter tops, lighting fixtures, appliances, plumbing fixtures, door and bath hardware, and landscaping. The list can go on, depending on the whims of a builder or architect. Even the brick or stone on a house's exterior could be governed by an allowance.

A good number of homeowners blindly assume that the builder has read their minds for the quality of these finishes, and priced the contract accordingly. Unless you can walk into a completed project to use as a basis for comparison, signing a contract laden with allowances is a recipe for busting the budget. Ask questions first to see what specific materials their stated allowance will furnish. It is better to err on the high side, for whatever part of an allowance you do not use should be returned as a credit.

Whether you are working with an architect or a builder, I encourage you to put forth the effort to choose as many material selections as possible early

in the process. This is one of the best ways to fulfill our six-step process of controlling the cost and fulfilling your expectations. To be sure, you may be required to do a lot of running around in a relatively short time, but I guarantee positive results.

Balancing the Budget

The final concept of this chapter is balancing your requirements for "space" or area with material finishes. We'll start with an experience I had on a new home project many years ago. A young couple, who both apparently came from very affluent families, interviewed me to design their dream house. They had already purchased a lot in a very upscale golf course development and had expensive ideas about their project. For the better part of two hours, they described a 6,000-square-foot house with the best in material finishes and equipment. When the subject of their budget finally entered the conversation, I held back my laughter when they stated a figure well below the cost required to turn their ideas into reality. We went back and forth discussing the relative costs of building the shell and finishing the interiors. They refused to reduce the large area, stating the house could not be less than 6,000 square feet. In frustration, I finally asked why they would want such a large house when their budget would force them to use cheap finishes. As an example, I named a well-known, low-cost, local kitchen cabinet company that frequently advertised on television. Seconds later, the young lady whacked me across the leg and informed me her father owned that particular company! As I struggled to remove my foot from my mouth, I figured this architectural commission was long gone.

Actually, I did get the job, and the couple and I proceeded to carefully analyze their space and finish requirements to establish a means of balancing their budget. At certain points during the design of every custom project, homeowners find themselves in the candy store with eyes bigger than their pocketbook. This is the spot where projects usually become derailed. Accept the fact that this may occur and have the following strategies ready to use.

When faced with budget imbalances caused by a design program larger than your budget, learn to prioritize and compromise. During the design phase in chapter 8, I encouraged you to first be loose in creating your program. Now the time has come to tighten up and make the tough decisions. Your goal should be to get the "biggest bang for the buck."

Project cost consists of two components. The first is the building "shell or envelope," which includes the site work, structure, exterior finish and utility systems, and the second is the interior finish package. The design process usually determines the exterior form and the area first, accounting for perhaps 60 percent of the total cost. Only after the establishment of the shell design is the cost of interior finishes considered. This time split in determining costs is an obstacle that you must overcome. How do you know how big to design the project without knowing the interior finish cost portion? I recommend that your written design program also include the level of interior finishes you want. Your goal is to reserve enough project funds for the interior finishes by controlling the area early in the process.

Let's say you want to build a 3,500-square-foot house. You can afford this size house only if you use inexpensive finishes: low grades of carpet, cabinetry, appliances, windows and so on. On the other hand, if you feel you must have lots of hardwood floors, two fireplaces, and a marble master bath, you must scale the area down to 3,000 square feet. Exterior materials will also have an impact, as brick costs more than siding, wood roofing shakes are more expensive than asphalt shingles, and windows are triple the price of solid walls.

As the design evolves into selecting the finishes, have your building professionals prepare a cost estimate for the entire project. Figure 9-2 illustrates a preliminary construction budget prepared by an archi-

tect. Professionals will furnish their estimates in many different levels of detail. I always try to present them on a trade-by-trade basis, listing each subcontractor's cost. This is the most comprehensive method, as you can visualize where the costs are concentrated. Note that at the bottom of the estimate, entries are made for the general contractor's overhead and profit.

Continual updates of these estimates are of great benefit as the design develops. By comparing different subcontractor values from previous budgets, you can track how your decisions have affected the project cost. Did you recently increase the area by 300 square feet? When you visited the appliance showroom did that stainless steel range suddenly become a must? Each of these items should be reflected in succeeding estimates. Require updates after each set of design revisions. They will enable you to monitor project costs more closely and avoid going overboard with too many goodies.

Examining figure 9-2, you will notice that a contingency or "fudge factor" is included in the preliminary cost estimate. Especially at the beginning of the design process, this safeguard is important because all of your requirements have not been identified or budgeted. This acknowledges that items still remaining to be selected will invariably require additional funds. The fudge factor should be larger when you start designing and can be reduced as you reach the conclusion. The surplus stash of funds will nearly always be used as you are exposed to all the possibilities of space and finishes. If you're a homeowner with iron discipline, there may be enough left at the end to keep in your pocket. In the event your reserve funds are not used, reward yourself with a night on the town!

Returning to our example estimate in figure 9-2, you can see that the preliminary budget has already been exceeded. On the majority of custom projects I've been involved in, the initial anticipated budget is too

Richard Preves & Associates, P.C.

Architecture   Planning

Figure 9-2

Preliminary Construction Budget

Client Residence
November 20, 2000

Note: An architectural opinion during preliminary design will vary as much as 10-15% from the final cost.
* - Allowances to be discussed.

1.	Permits - Allow	$ 6,000
2.	Site Clearing, Excavation, Grading Allow	11,000
3.	Concrete	26,000
4.	Steel	1,500
5.	Rough Carpentry	87,000
6.	Roofing and Sheet Metal	10,500
7.	Windows and Exterior Doors	26,000
8.	Fireplace	4,000
9.	Insulation	3,000
10.	Driveway, and Culvert Pipes - Allow	10,000
11.	Garage Doors and Openers	3,000
12.	Gypsum Board	12,250
13.	Electrical	24,000
14.	Plumbing, Gas Piping	20,000
15.	Heating and Air Conditioning	19,000
16.	Well	6,000
17.	Septic	9,000
18.	* Cabinetry, Countertops and Built-Ins Allow	30,000
19.	* Custom Stairs and Railings - Allow	8,500
20.	* Ceramic Tile – Allow	6,250
21.	Wood Flooring @ $ 7.00/ Sq. Ft.	10,000
22.	Carpeting – 230 yards @ $ 25.00/yards	5,750
23.	* Appliances (All New) - Allow	8,000
24.	Shower Doors, Mirrors and Bath Accessories	3,000
25.	Painting	18,000
26.	Trim Carpentry	17,500
27.	Interior Doors and Hardware	13,000

Sub-Total = $ 398,250
General Contractor's Overhead and Fee @ 15% =   59,737

Sub-Total = $ 457,987
3% Homeowner's Contingency   13,740

Total = $ 471,727

@ 3,500 SF = $ 130.85 SF

ORIGINAL PROJECT BUDGET $ 450,000
OVER BUDGET   ($ 21,727)

low to fulfill an evolving design program. While at first it is a good plan to develop as many design ideas as possible, the resulting cost consequences can quickly bring you back down to earth. At the point where everyone realizes he cannot afford everything he wants, a session with the builder or architect is in order. With design goals and available funds starting to head in opposite directions, compromise is required to keep the budget in balance.

This adjustment process I call *prioritizing* is the key to a successful project. Take your design wishes and organize them on paper from the most important at the top to the least significant at the bottom. Working with your professionals, have them estimate the cost for each item, as shown in our example in figure 9-3. An additional bedroom will cost a certain amount. The hot pink whirlpool will run you this much more. By prioritizing your goals, you can start adding up the associated costs until you have reached your budget limit. Draw a line at this point and forget the rest! Even homeowners with very ample budgets must organize their preferences.

Recalling my story about the young couple, you will invariably have to balance the size of the project versus the special details and finishes. For example, is an extra bedroom or larger bedrooms more important than amenities such as fireplaces, hardwood floors, or a home entertainment system? These financial choices are the tough decisions every homeowner must make.

Use your professionals as a resource throughout this prioritizing process. From their experience, they can help guide you into making the right decisions. Often, I have steered my clients away from spending additional funds on items that at first sound like a good idea. Extra fireplaces are rarely used after the first couple of years, skylights in bedrooms act like an alarm clock during early summer mornings, and that perennial favorite, the indoor grill, never works

very well! Carefully consider whether you will really use every item you think you want. Gadgets and gizmos soon loose their luster if they are never used.

Plan for Future Expansion

You may want to consider another suggestion I offer to clients with limited budgets. If you have expectations of having more money to spend on your project in future years, plan for altering the design after the project is first constructed. Space and amenities are the primary targets. Anticipate an addition down the road by planning it up-front. Hoping to add more bedrooms or a larger family room? Have your designer include the necessary elements now so the extra space can be effortlessly added. Preplanning an addition will be cost effective because you won't have to alter much of the house to accommodate the future new space.

If your original space requirements will not permit you to build that gourmet kitchen you would love, no problem! Design the current kitchen with enough area to accommodate your future needs but install less expensive "disposable" cabinetry and appliances. When you have the funds to finally remodel the kitchen, throwing away the cheaper cabinets and appliances will not be too painful! Visualize your project as a work in progress, making long range plans for future modifications. It's a great way to live within your budget and yet still have ambitious plans to dream about.

The design development step is all about squeezing the most value out of your dollar. If you have been skilled in finding a resourceful project team, its members can show you innovative ways of making space look larger than it really is or finishes that look like the real thing but cost far less. There are a lot of "tricks of the trade" to consider. Spend the time to explore the possibilities. It will require some effort, but you will enjoy the fruits of your labor.

Richard Preves & Associates, P.C.

Architecture   Planning

Figure 9-3

Design Priorities

Items in Order of Importance
January 10, 2001

		Additional Cost
1.	Fifth bedroom	$ 20,000
2.	Full finished basement instead of partial	8,500
3.	Fourth car garage	13,500
4.	Brick exterior instead of wood siding	<u>15,500</u>
	Total Extra Cost Accepted	$ 57,500
5.	Wood shake roof instead of asphalt shingles	7,000
6.	Custom wood stairs instead of stock	2,750
7.	Oversized whirlpool tub in master bath	1,500
8.	Closet storage system instead of rod and shelf	1,900
9.	Upscale kitchen cook top	1,000
10.	Heated towel bars	850
11.	Solid wood doors	3,500
12.	Wet bar for six	3,500
13.	Built-in wine storage cabinet	2,000
14.	Home theater entertainment system	4,500

Chapter Nine Recap

- If you decide to use an interior designer, enlist his help early rather than later in the design process. Coordinate the responsibilities of the architect and interior designer so there are no overlaps or holes in providing the overall project design.

- When faced with budget imbalances created by your desire for more than you can afford, learn to prioritize and compromise. You are juggling two separate balls, the amount of floor area in your house versus the cost of construction materials.

- Selecting all construction materials or "finishes" before the plans are complete will help you control the budget.

- Working with your professionals, verify with our checklist that all finishes have been selected. Use catalogs, or even better, visit showrooms or other completed projects, to make informed decisions.

- Have your professionals provide frequent budget updates as the design process evolves to monitor construction cost. Build-in a "fudge-factor" to cover unanticipated expenses.

- Know what a *"cost allowance"* means. Often builders may suggest using dollar amounts to cover finishes not yet selected. If the allowance is below your design expectations, the amount to cover the material upgrade will be coming out of your pocket. Verify that allowance amounts are adequate.

- Prepare a list by ranking from top to bottom each of your indulgences beyond the basic necessities. Have your professionals furnish a dollar amount for each item. Starting with the top priority, work down the list until your budget amount is reached.

Chapter 10

Prepare for Successful Project Bidding

I was once asked by a friend to consult with his sister and her husband, who were building a new custom home. The couple explained that they hired an architect to prepare the plans and had just received general contractor bids. The mystery that required my assistance was determining why the bids were so different from each other. Of the three bids submitted, two were 25 percent apart and the third actually had no dollar amount, but mentioned several percentages of construction cost. With construction proposals so far apart, they had no idea whom to hire to build their project.

I first reviewed the contractor bids and then asked to see the architectural drawings. The drawing set contained just a few scanty sheets. Floor plans and the exterior of the house were shown, but little detail had been included for the interior finishes. No instructions regarding bidding requirements for the general contractors were to be found. "The problem is unfortunately very simple," I told them. "The plans are incomplete and the generals had to guess your intentions. Apparently, each bidder guessed differently."

The bid with no construction numbers reflected the problem of insufficient information. This bidder proposed to keep a running account of what would be

spent and add a percentage to cover his expenses and profit. "Why would anyone want to build a house in this manner?" they asked. I advised them to meet with the architect and the contractors to determine what missing information should be added to the drawings. Once the plans were revised, the contractors could refigure their bids.

We have come across yet another issue that often creates many unhappy stories. If construction bids vary in their pricing, should the homeowner take his chances with the lowest bidder or accept the highest bid, hoping that it will include more of their expectations? Even the odds at the Las Vegas slot machines are better than this wager!

This chapter will introduce you to several concepts to insure your contractor bids will be complete, accurate and competitive. Everything we have previously discussed has been a prologue to reaching this point in the project process. The rubber is now meeting the road. When the sawdust starts flying, there is no turning back! Good planning will always pay off; poor preparation guarantees many storm clouds on the horizon.

Contractor Bidding Options

First, you must understand the different approaches general contractors and builders can use in charging for their work on your project. All methods produce the same finished project, but the difference determines how the final price is calculated. Referring to figure 10-1 you will see these three different bidding approaches: *Lump Sum, Cost Plus*, and *Open Book*.

Any one of these methods can be used with either the Traditional or Design/Build project delivery system, and any of these methods can be adapted to your project. Each method has its own advantages and disadvantages when it comes to timing and pricing. For example, some custom home projects have special circumstances, such as tight time schedules, unique design issues or homeowners who can't make quick decisions but need their project finished quickly. Nailing down a complete project cost before construction can be difficult for these reasons. By using the Cost Plus billing method, a general contractor or builder can begin building without necessarily knowing the total price.

It's vital that you understand the advantages and disadvantages of each bidding method. Used correctly, they can produce good results and offer alternative paths to finishing your project on time. If used incorrectly, some of the methods can be dangerous to your financial health.

The most common approach to determining construction cost in the building industry is the Lump Sum bid. In this scenario, the builder or general contractor will use the architectural drawings and specifications to generate a guaranteed total price for constructing the project. The builder in the Design/Build System will produce a lump sum cost during completion of the drawings. Since he is involved during the design process, he can get a headstart on determining the final price. In the Traditional System, the general contractor does not become involved until the architect has completed the construction documents.

Lump Sum bidding's first advantage is the safety and comfort of knowing the total project price up-

front. The design and details contained in your set of drawings will cost a certain amount to build. Ideally, if no changes are made to the drawings, the construction cost should not change. A Lump Sum proposal can simply state the following, "ABC Company will furnish all materials and labor to construct the project in accordance with the drawings prepared by XYZ Architects." Nice and simple, right? That is why the great majority of construction contracts employ this method. It is also easy to understand and administrate for periodic payments to the contractor.

Another advantage of this concept is that it provides a good basis for competitive general contractor bidding. Distributing your drawings to general contractors enables you to receive and evaluate multiple bids based upon the same information. As I've mentioned before, when I was a general contractor, I knew I must "sharpen my pencil" when competing for a project against other bidders. We have already discussed the importance of competition and its value in producing the lowest price. Construction is the final and most important point in the process where competitive bidding should be applied. There is no better feeling than knowing you are getting the best deal! The numbers are in, you know the final cost, and the rest of the project should be a snap.

Alas, as with all things in life, nothing worthwhile is that easy! There are a few traps in the Lump Sum method. First and foremost, this guaranteed price you just received is only as dependable and complete as the drawings it is based on. From our discussion in chapter 3, you'll remember that not all architects and their services are created equal. Depending on the services and expertise you paid for, the drawings produced by different architects will vary in their level of completion.

Recall the opening story of this chapter. Architectural construction documents that are complete in detail and cover all the bases will produce correspondingly comprehensive bids. Drawings that are

Figure 10-1

Construction Bidding Options

LUMP SUM BID	COST PLUS	OPEN BOOK
• Guaranteed maximum cost	• Contractor fixed fee on top of sub-contractor and direct costs	• Similar to *Cost Plus*, but contractor will show owner all sub-contractor bids
• Basis for competitive bidding	• Fastest method to build project	• Guarantees competitive bidding at subcontractor level
• Cost based upon content of drawings	• Final cost not always established until project's end	• Difficult to find builders/contractors willing to work within this system
	• Lack of competition	
	• Use Guaranteed Maximum Price (GMP)	
	• Competitively bid overhead and profit fixed fees	

incomplete will likely yield deficient construction proposals. As you sit back, basking in the glory of bringing your project in on budget, a surprise is in the making as you discover during construction that the carpeting specification was inadvertently omitted from the plans and the contractor assumed you were furnishing it separately. This is the same scenario we discussed in chapter 2. Lousy drawings always generate lousy bids!

Lack of steadfastness in your own design decisions can also play a major role in generating budget busters. Lump Sum bid guarantees do not extend to construction changes requested by the homeowner. When your friends moan about how much over budget their project was, ask them how many changes they made during the building phase! This is by far the most frequent cause of busted budgets. If you stick to your guns and stay with your original decisions, the original bid will be close to your final cost. My ideal client would leave town at the start of construction and only return when we can turn over the keys. This way they cannot have second thoughts about their original design. Don't become your own worst enemy, stay the course!

The Cost Plus method is our second bidding option. Here things are going to become a little confusing. A builder or general contractor using this method will pass along to the homeowner all direct construction costs consisting of materials and labor, such as carpentry, plumbing and electrical work. A predetermined fee based upon a percentage of the final cost is added to cover the contractor's overhead and profit. This concept is actually borrowed from commercial construction where it is used more frequently. It should be employed only when special project circumstances dictate the need for an alternative bidding option.

One special circumstance usually involves time. If your project needs to be built quickly and must proceed before all your design decisions are complete, this option can produce results. While the builder or general contractor is pouring the foundation and erecting the exterior framing, you are completing your interior finish selections. Instead of awaiting the completion of the architectural plans, construction begins earlier, and saves considerable time. As soon as the design basics can be shown on a plan, a building permit can be obtained. This process is also know in the commercial sector as *Fast Tracking,* for obvious reasons. If you need space quickly, Cost Plus can produce results.

In exchange for speed, certain circumstances can put you at a disadvantage. The first concerns total project cost. Until the project is close to completion, the final price is unknown. You have created a moving target similar to running a tab for a party at the local watering hole. After you finish for the night, the bartender totals your drinks and hands you the damages! Actually it is not quite this bad, because budget limits can be established at the beginning of the project, using numbers based on past experience.

If the budget estimate is set close to the expense level of the finishes you'll choose later, cost overruns should be minimal. A budget set too low for any reason, whether through honest mistakes or intentionally, will produce disastrous consequences. This is very much like Russian roulette. You are trusting your builder or general contractor not to hand you a loaded gun!

If you decide to use Cost Plus to save time, use your common sense. During the budgeting process, ask your professionals to explain their assumptions in establishing the cost level for each material and finish. Expecting wood floors in your family room? Verify that the flooring line item contains enough funds to cover the expense. Go down our list from chapter 9 to double-check each selection. Cost Plus is not for the lazy homeowner; using this method means you have just signed up for extra work.

The second disadvantage of this method is lack of competition. Competitive bidding has been virtually eliminated. Since you must choose a builder or general contractor to start construction quickly, you are trusting him to be honest and fair. He is under no obligation to find the lowest cost for every trade because he isn't competing with other builders and contractors to get the job. Once you give the green light for construction to begin, the meter is running. This is very similar to a taxi ride. You know the cab driver will get you to your destination, you are just not sure whether he'll take the shortest route or build up the fare by going the long way!

To combat this potential problem, there are two techniques to alleviate some of the risk. Before you finalize your deal, request your chosen professional to work within a *Guaranteed Maximum Price* (GMP). This concept requires that given a generally predetermined design program, the Cost Plus price method is guaranteed not to exceed this figure. Many professionals will artificially inflate this number beyond the real price to protect their bottom line, but it can afford you some comfort of knowing the upper limit. Insist on a GMP before signing the contract.

The second technique involves competitively bidding the overhead and profit portion of the builder or general contractor's cost. Since this portion of the budget is usually based on a percentage of the direct cost, request each candidate to quote their fee. For example, three general contractors, asked to submit fee quotes, could return with proposals at 12 percent, 15 percent and 17 percent of total construction cost, respectively. This means that after all costs for materials, equipment and labor are totaled, the selected contractor's preset fee is added to make up the total project cost. If the fee is established at 12 percent, and the direct costs are $100,000, the total cost equals $112,000. Contractors will take this deal any day, because their overhead and profit level is guaranteed, eliminating any risk. Requiring fee proposals introduces a degree of competition into the Cost Plus method. These fees can also be negotiated between competing contractors to lower your cost.

However, keep in mind that if you have the time to adequately design and bid your project for a Lump Sum, reject any builder or contractor who insists on using Cost Plus. You would be taking on too much risk with no chance of reward. Lump Sum is by far the safest way to go.

Taking the competitive concept one step further in the Cost Plus option, we reach the third bidding choice, *Open Book*, an extension of Cost Plus, which also comes from commercial construction. Open Book expands the bidding of builders' and contractors' fees to include the subcontractor level. The builder or contractor is required to "open his books" for your review and reveal the cost of each trades' work. A minimum of three bids for each major subcontractor is submitted for your analysis. In this case, for example, you will see three electrical proposals, ranging in price from low to high. By examining each trade's set of bids, you can verify that the lowest price was used.

When you combine the bidding for builder's and contractor's fees with the Open Book method, you have put competition back into the process. But expect to see multiple proposals for the major trades only. One bid is usually considered sufficient for minor trades, such as fireplace doors, garage doors, or shower doors.

The prime disadvantage of the Open Book method is that few residential builders or contractors are willing to be this accommodating. Besides requiring quite a bit of additional work on their part, Open Book requires builders and contractors to reveal many of their costs, and most prefer to keep their numbers close to the vest. Don't anticipate using Open Book on smaller projects. Most professionals will not deem the lower profit worth the extra effort. Larger projects are more appropriate for this method. Plan on committing more of your time to utilize this technique. Again, only special or unique circumstances regarding project timing makes this method worthwhile.

Architectural Drawings; the Complete Package

We've reached another critical fork in the road. With a set of comprehensive documents, you will be on the path toward your final destination of project success. Keep in mind, though, that a road filled with bumps and hairpin turns awaits you if you must use incomplete drawings as your vehicle. In previous chapters, I've mentioned the importance of complete architectural construction documents for accurate bidding. Now you're probably wondering how a typical homeowner, inexperienced in reading the trade language contained on drawings, can make sure all the required information is included in the architect's plans.

First let's review the importance of these large sheets of paper containing lots of lines and text. I'm convinced that the time it takes to assure that drawings and specifications are complete is time well spent. Drawings and specifications—the contract documents—are the most crucial elements for full communication between you, your architect, and the contractor or builder. Contract documents are vital for obtaining construction bids and building permits, as well as for building or remodeling your house. All of the written decisions, goals, and requirements of your design elements should be included in the drawings and specifications.

This set of documents specifically describes the quality, size, and quantity of every piece of construction to a contractor or builder. They are the culmination of the design process. The drawings will have graphic representations and listings of what is to be included. The specifications are a set of written instructions on the quantity and quality of the materials that are built into the project. It is critical that all materials selected during design development be shown for complete bidding. Clear, complete contract documents put you in charge of the project, keep costs down, and minimize extras.

There is a direct correlation between the amount of detail on the architect's documents and the completeness of the contractor's construction price. Contractors and their subcontractors are not mind readers. It's not a good idea to depend on the contractor to cover any lack of design detail with adequate funds within their bid. If the documents do not have sufficient information on the drawings and specifications, the contractors will either make their own assumptions how something is to be done, or they may exclude the work completely. They could insert an allowance into their bid that may be well below the level of your expectations.

These drawings and specifications are also a part of the legal construction contract that we previously discussed. If items are not contained within the documents, there may be no legal requirement for the contractor to furnish them. This is where horror stories always start to evolve. As we remarked in the discussion of the Lump Sum Bid option, omitted items often create cost extras. Legally, the only scope of construction is shown on the drawings.

To familiarize you with these documents, I have included some examples of drawings that have been reduced in size to fit the book's format. While you may not be able to read the notes on the drawings, they will make you aware of how a project is organized for bidding and construction. As we review each plan, you will notice the amount of detail contained in a typical set of drawings and specifications. Get ready now for the introduction to our very abbreviated short course, Blueprint Reading 101.

Design professionals vary in their use of written specifications. (See figure 10-9.) Information not specifically contained within the drawings are referenced in a written section. This set of instructions provides construction professionals with material manufacturers' stock numbers and colors. For instance, if a bathroom elevation calls for ceramic tile, the specifications will list a corresponding entry: "American

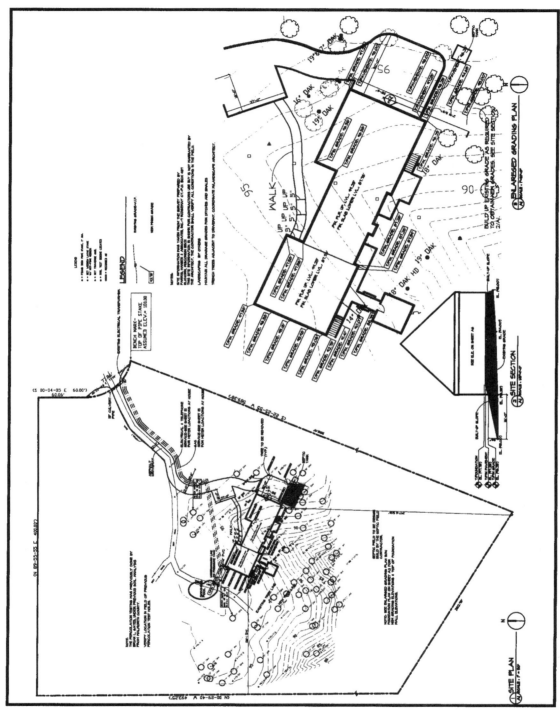

Figure 10-2
Site Plan

Your new house or addition is shown on a drawing that illustrates the physical dimensions of your lot and indicates the exact placement of the structure. Accessory features such as driveways, exterior decks or patios and utility connections are also depicted. Basic information relating to grading, or the sloping of the ground around the new construction, should be included. Setback lines, or the minimum distance required by local zoning codes for construction within your property is another required element. Normally, this plan is not required for remodeling projects that do not affect the exterior.

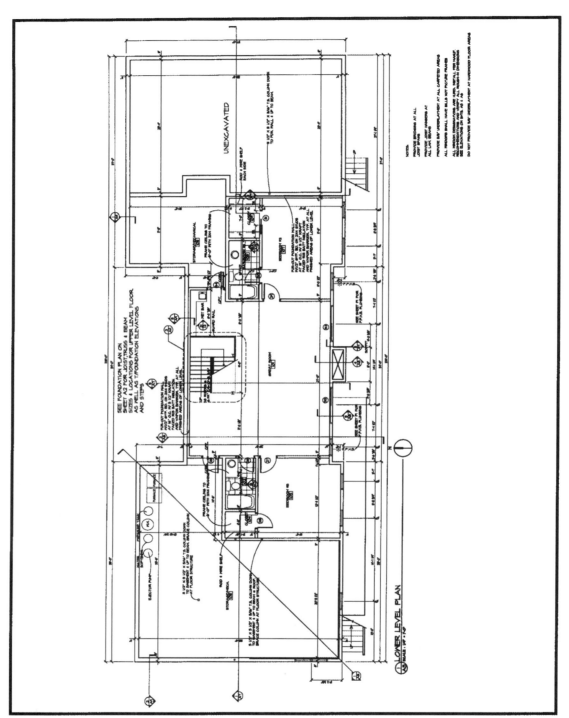

Figure 10-3
Floor Plan

Each floor or level is shown as if you were looking down from above. The first floor would be seen from a position below the second floor. The second floor is illustrated from a viewpoint under the roof. All rooms with their respective sizes, walls, windows, and doors should be specifically identified and dimensioned. Kitchen and bathroom equipment and fixtures are also included. Electrical lighting, power outlets and telephone jacks throughout each floor and other important elements are shown on a separate plan.

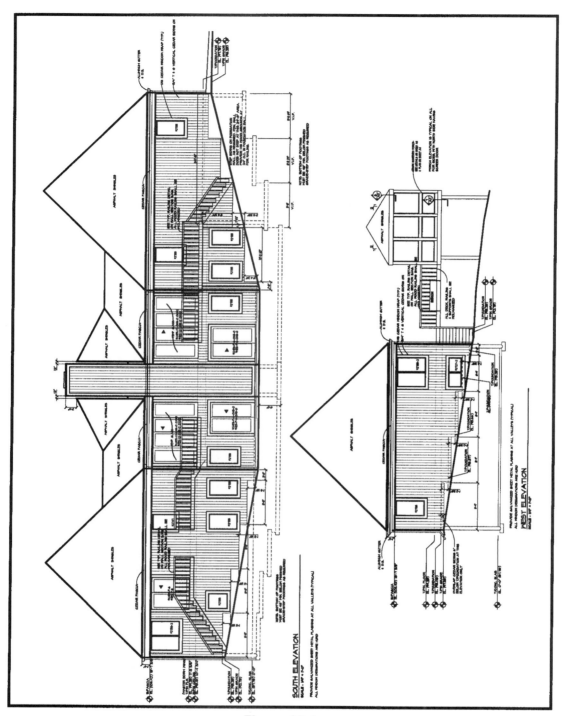

Figure 10-4
Elevations

Exterior views of each side of the project are exhibited as if you took a photograph looking directly straight-on at each side. Elevations, or exterior views of projects that are simple rectangular boxes are fairly easy to understand. Structures that have many projections or angles will be distorted when flattened on the drawing. The models, CADD renderings or pencil sketches we discussed in chapter 8, can help you to visualize this change from three-dimensions into the flat, two-dimensional language of contract documents. It is a source of pride among professionals and tradesmen that any project can be understood and built from two-dimensional drawings, no matter how convoluted and complex the design! Exterior materials such as siding, masonry and roofing should also be identified.

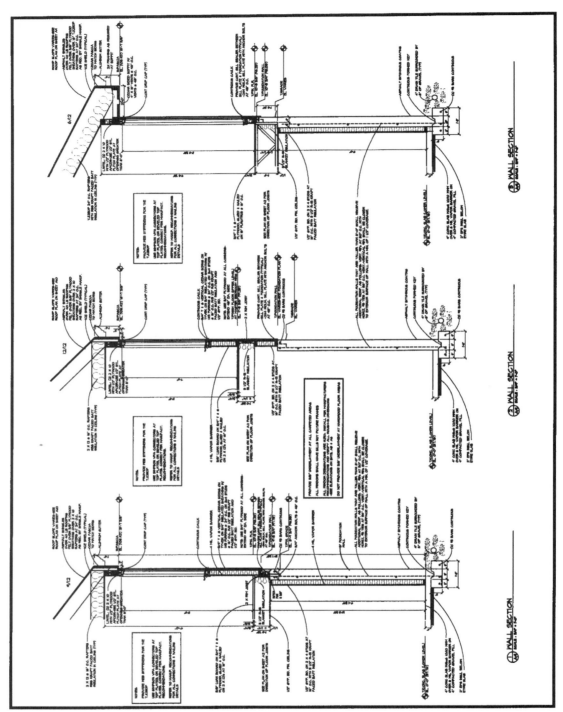

Figure 10-5
Sections

Here the going gets much tougher for the average homeowner. Imagine cutting a grapefruit in half and looking at the inside straight-on. This is the concept of drawing sections. We take different areas of the house and cut them apart to illustrate how the nuts and bolts are assembled. Common sections are shown through exterior walls, illustrating the connections between foundation, outside walls and the roof. Specific information regarding the size and placement of the structural frame are contained within these sections. For more complex projects, sections cut through the entire house are sometimes used to better explain the relationships between different components.

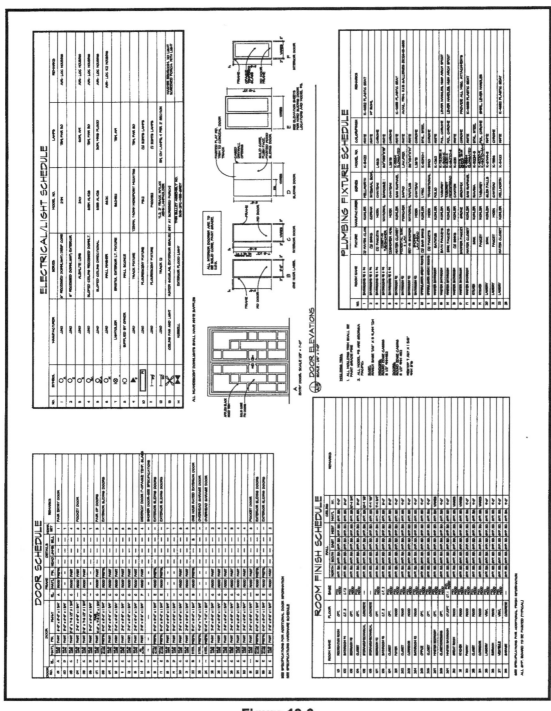

Figure 10-6
Schedules

I like to use simple tables in my drawings to itemize interior finishes, plumbing fixtures, lighting fixtures and doors. These "schedules," as they are called, provide a comprehensive list that is easily referenced. For interior finishes, a Room Finish Schedule is prepared that lists every room with its respective floor, wall, ceiling, and baseboard materials. This schedule can be refined to reflect the paint color of each surface. The same principle applies to describing doors and hardware, plumbing, and lighting fixtures. Schedules provide a well organized, convenient checklist for use by contractors and material suppliers, as well as making it easier for the homeowner to verify their selection of finishes.

129

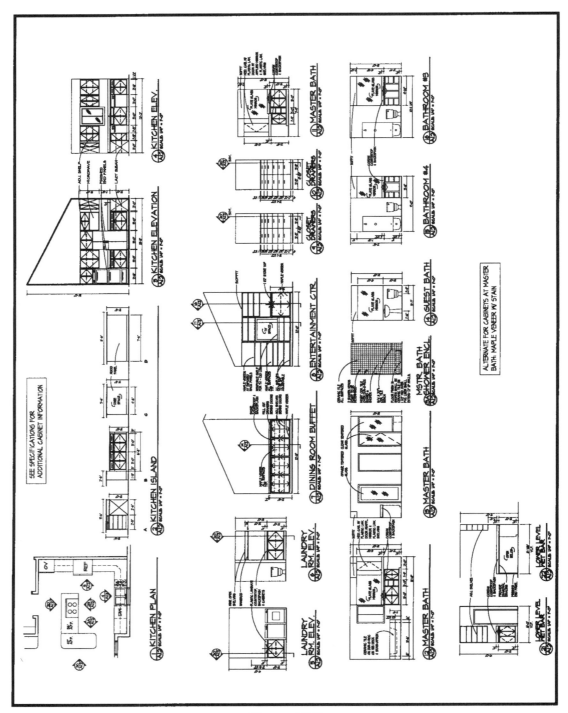

Figure 10-7
Interior Elevations

Just as the exterior of the project is illustrated on elevations, so too are parts of the interior. Kitchen, bathrooms, fireplaces, and built-in cabinetry are drawn two dimensionally, describing their appearance. As an example, kitchen elevations would explain the size and placement of cabinets, appliances and accessories. Bathroom designs depicting fixtures and cabinetry placement, as well as ceramic tile on walls, give the tradesmen specific instructions beyond what can be indicated on floor plans.

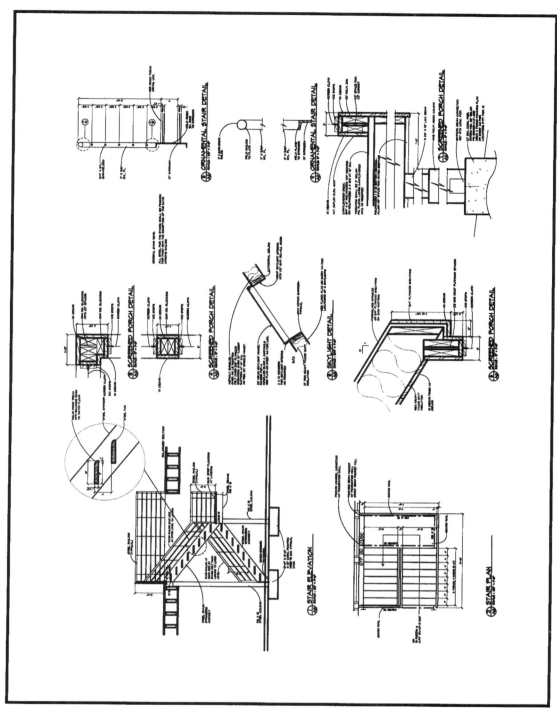

Figure 10-8
Details

Special construction conditions requiring specific instructions are provided in enlarged drawings called Details. *This category of drawings would include decorative trims, door and window moldings, stairs, railings and unusual design conditions requiring further explanation. The number of details included in a drawing set varies with the complexity of the design. Straightforward projects built with the basics do not require much additional detailing. Houses that contain higher levels of materials and custom assemblies correspondingly call for an increased number of details.*

Figure 10-9

Specifications

1.   **General Conditions**
American Institute of Architects (AIA) A201 latest edition shall serve as the basis for general conditions for this project.

2.   **Special Conditions**
A.   Conform to all local and national building and zoning codes adopted by Anywhere County.
B.   The General Contractor shall verify all grades, lines, levels, and dimensions as shown on the drawings, and he shall investigate thoroughly all existing site conditions. If an error or omission appears in the plans or specifications, the General Contractor shall, before proceeding with the work, notify the Architect in writing of each error and omission before proceeding.
C.   The General Contractor shall provide all foundation surveys for the project as required by Anywhere County.
D.   The General Contractor shall provide the Architect shop drawings or submittals for review before installation.
E.   The General Contractor shall apply for, obtain and pay for **all** required permits.

3.   **Site Work**
A.   The General Contractor shall limit his construction activities to areas indicated on the site plan. The General Contractor shall be liable for damage outside the contract limit area.
B.   Rough grading shall be done with clean fill free of debris or other material.
C.   Finish grading shall be by this Contractor. Use 2" – 3" black dirt free of debris.
D.   Coordinate all incoming well, septic, electric and gas services.
E.   Verify setbacks on the site prior to start of work.
F.   Erect temporary fencing for the septic field in accordance with Anywhere County Health Department regulations prior to commencement of any work.
G.   The Contractor shall supply and install all site erosion control as required by Anywhere County.
H.   The Contractor shall remove trees and brush only as required for construction. Coordinate with the Owner and Architect.
I.   Asphalt Driveway: Driveway shall be 2" minimum thickness over 6" compacted stone base. Provide seal-coat 90 days after installation.

4.   **Concrete**
A.   All concrete shall be minimum 3,000 PSI, 5-1/2 bag mix.
B.   Reinforcing steel shall be 6,000 PSI yield strength.
C.   Concrete work shall be placed when ambient temperature is 40°F for 24 hours.

5.   **Steel**
A.   Steel shall be A36, shop primed. Provide shapes as indicated on the drawings.

6. **Rough Carpentry**
   A. Framing lumber except as indicated on the drawings shall be pine or fir #2 grade, or better.
   B. Provide laminated beams and steel connections as indicated on the drawings.
   C. Cedar siding shall be 1x6 small tight knots, kiln dried, tongue and groove vertical #2 cedar or better, smooth side exposed. All cedar shall be installed with non-corrosive 3" long nails in straight level lines. Installation of nails in a random pattern or off level shall be cause for rejection.
   D. All sealants shall be silicone type.

7. **Dampproofing**
   A. Dampproofing around the foundation exterior walls shall be sprayed asphaltic dampproofing. Do not spray above areas exposed above grade.

8. **Insulation**
   A. Building Insulation shall be fiberglass batt insulation. Furnish insulation to conform with R values as indicated on the drawings.
   B. Provide 4 mil poly vapor barrier.

9. **Roofing and Gutters**
   A. Roofing Shingles: Tamko™ Organic Seal-Down 25, 3 tab asphalt shingles, color: weathered wood. Install per manufacturer's recommendations.
   B. Provide sheet metal flashing at all areas required on the drawings, including plumbing, heating and chimney penetrations.
   C. Gutters and downspouts shall be seamless aluminum, standard color.

10. **Windows, Exterior Doors**
    A. Provide Pella™ vinyl clad manufactured windows as scheduled on the drawings. Color to be Sand. Provide jamb extensions and window screens.
    B. Provide millwork window sills, not "picture frame type", unless noted on the drawings.
    C. Provide insulated glass, Low E Argon type.
    D. All fixed windows shall not have operating hardware.

11. **Garage Doors**
    A. Garage doors shall be flush style hardboard with vertical 1x6 tongue and groove cedar siding applied on the exterior. Provide and install electric operators and push button wall and remote controls, and exterior digital combination keypad for each door. Sizes shall be as indicated on the drawings.

12. **Gypsum Wallboard System**
    A. Install ½" Gypsum wallboard as indicated on the drawings. Provide all corner beads at all gypsum board corners. Gypsum board finishing shall consist of 3 coats mud and sanding. Comply with all requirements of Gypsum Association GA-216.

13. **Doors**
    A. Furnish and install all doors, closet shelving, etc. as shown on the drawings. Doors shall be solid core, paint grade. Front door shall be fir for staining.
    B. Furnish and install all moldings, casing, and base as scheduled on the drawings.

Olean™ 2 in. x 2 in. ceramic mosaic, Hot Pink, number 6879 with white grout." Certain designers will place this information right on the plan or elevation. Others prefer to keep the information organized within the specification section. Either way is acceptable, as long as the requisite information appears somewhere within the drawing set.

Specifications are also used to set standards for workmanship. As an example, instead of including ten pages of instructions on the proper method to install drywall, I will simply insert a phrase that requires the contractor to conform with the recommendations of the Gypsum Board Association. With this simple reference, I tell the contractor he is required to adhere to all requirements of a recognized regulatory organization described in their comprehensive documents. Nearly every family of building materials has a written standard for workmanship, which can be employed as a reference for your project.

We have now reached a point where problems often crop up. How much information regarding installation and assembly of materials should be included within a set of construction documents? Many professionals depend on the concept of *common knowledge*. Architects assume that a contractor or builder and their subcontractors will know how to properly lay brick, hang a door, install windows or paint a wall. Here residential project drawings depart from their commercial cousins. Large-scale building projects such as offices, shopping centers, and schools actually have specification books describing these requirements that can run into the hundreds of pages.

This is rarely provided on house projects. Deciding on the amount of information to include in specifications is a judgement call by the architect. Comprehensive instructions for workmanship included in your documents will furnish an extra layer of insurance toward achieving good craftsmanship. If you feel this safeguard is necessary on your project, be prepared to pay additional fees. Their preparation

and inclusion takes the architect significantly more time.

If you are using the Traditional System, your design professional should prepare a set of instructions for the general contractors. (See figure 10-10.) This short document sets the ground rules for the type of bid to be submitted, such as the Lump Sum method. A date establishing a bidding deadline has been indicated in our example. Other information specific to the nature of the project has been included. These instructions should be combined with the contract documents sent to each contractor for bidding.

Now that you have been inundated with all this technical razzle-dazzle, how is the novice homeowner to know if the project plans have been adequately prepared? This complex issue should first revert to the interviewing process. As an architect, I am proud of the amount of detail my office includes on our construction documents. I make it a point of showing prospective clients our drawings to assure them of our commitment to thoroughness. You should ask to view each candidate's documents as part of your evaluation criteria. Ask each professional to describe his approach to assembling a typical set of drawings. As you compare each firm's product, you will definitely notice a difference in their methods. As a rule of thumb, bigger is better when it comes to the size of a set of drawings.

Finally, at the time your architect or builder declares that the documents are ready for bidding or construction, request a meeting. Armed with your meeting minutes, telephone memos and material brochures, ask to be shown where each selection is referenced on the drawings. Verify that the specifics such as size, color, and finish, with manufacturer and model numbers, have been listed. Your building professional may take offense at this questioning of his or her competence, but stand your ground. I find this checkup mutually beneficial for me and my clients, as it results in a better job.

Richard Preves & Associates, P.C.

Architecture   Planning

Figure 10-10

Instruction to Bidders

Client Residence

1.   Use the enclosed bid form only.

2.   The bid is for one lump sum bid including all work and materials to fulfill the intent of the drawings and specifications. There is one bidding alternative for stone flooring.

3.   The contract documents for the bid consist of Sheets A1 – A12 and specifications, dated April 5, 2000.

4.   All work should be as specified. If you wish to propose a substitute, please advise our office during the bidding period for approval.

5.   All questions shall be answered in writing for the benefit of all bidders.

6.   The Contractor shall visit the site and familiarize himself with all conditions affecting the construction of the project.

7.   The owner reserves the right to reject any or all bids.

8.   Bid due date is Wednesday May 1, 2000. Bids should be sent to our office.

9.   The owner anticipates entering into an agreement with the successful bidder using American Institute of Architects A107 Contract, "Agreement Between Owner and Contractor".

10.   The successful bidder shall be required to furnish a current certificate of insurance prior to construction, including general liability, excess liability, motor vehicle liability, and workmen's compensation.

Figure 10-11

Proposal for Construction

Date: June 10, 2000

A. Project Name: Client Residence

B. Project Address: 1234 Main Street
Somewhere, US 12345

C. Submitting Contractor: ABC Contractors

D. Description of Work: Construction of new two story house

E. Architectural Drawings and Specifications Summary: A1 – A12 and Specifications contained therein, Dated April 15, 2000

F. Addendum: (Number and Date)
1. May 17, 2000
2. May 29, 2000
3.

G. Contract Amount: $ 469,645.00

H. Alternates Amount:
1. Stone floor installation Add $ 9,500.00
2.
3.

I. Allowances: $ 500.00 for Permits

J. Substitutions: Seal-Rite™ in lieu of Anderson™ windows

K. Time to complete contract: 270 (Calendar Days)

Authorized Signature:

For: _____

    ABC Contractors

Date:

    June 10, 2000

Analyzing Your Construction Bids

Even with the best architectural documents, general contractor proposals are often received with an unexpected spread of lump sum bids accompanied by allowances and exclusions. I am constantly surprised that despite specific instructions on the drawings, general contractors insist on using their own interpretations when quoting their price. An exact description of a lighting fixture could be included on the drawings only to see a bidder decide to substitute another fixture within his proposal.

Our goal is to insure that all construction proposals are compared on an "apples to apples" basis. Careful analysis by you and your design professionals is required to place all the construction players on a level playing field. Whether your bids come in with a 5 percent or 25 percent spread, a bid analysis is definitely recommended. Even bids whose numbers are nearly identical may change after evaluation. I have had several projects where the apparent low bidder lost the favored position after analysis revealed omissions that required their bid to be adjusted upward.

First, let me show you a typical general contractor proposal within the Traditional System. By scanning figure 10-11 you will see that the proposal contains a Lump Sum bid followed by several allowances and exclusions. Each bidder will have submitted different costs and separate modifications to their overall bid. To begin equalizing these discrepancies, request each contractor to submit a *Cost Breakdown*. Figure 10-12 illustrates a response from a bidder to this request. It should be organized on a trade-by-trade basis, each with an associated dollar value. If you have retained your architect to provide bidding analysis, this job is his responsibility.

After discussing the specifics of their proposal with each bidder, a *Bid Analysis* can be produced by your design professional. Figure 10-13 is an analysis compiled for two contractor bids. The original amount

for each subcontractor trade will be listed. Using the apparent low bidder's numbers as a base, upward and downward adjustments are made to compensate for each discrepancy. At the bottom of the form, the total adjustments for each bidder is tallied, yielding a final project cost.

By examining these subcontractor numbers and making a comparison between each proposal, imbalances may begin to appear. For example, matching the electrical number from bid A and bid B may reveal a difference of $5,537. Does this discrepancy come from separate labor, materials and profit expectations, or did the lower bidder forget an item? One possibility is that there was a lack of information on the drawings and the electrician plugged in a $1,500 guess that was $2,000 below the actual cost. Upon closer questioning of the general contractors, it turns out that this particular electrician omitted a special chandelier worth $1,231. There are countless other possibilities for explanations. Using these cost breakdowns can be the first step in identifying bidding problems.

Bidders can actually change ranking once this analysis is completed. Contractor proposals that were far apart can also suddenly tighten. A bid that was below the estimated cost was revealed as too good to be true because of several costly omissions. This process requires additional time by your architect but is worth the effort. I consider this task a basic service and normally include it in my full-service package.

Interviewing Contractors after Bidding

Before making your final selection and hiring a general contractor, an interview is a must. The following topics should be on your agenda. Meet with the finalists produced by the *Bid Analysis*. If there are one or more clear candidates whose adjusted numbers are within 5 percent, take the time to meet with each. Using our interviewing techniques and questions discussed in chapter 7, evaluate each contractor for the proper fit in joining the project team. Iden-

ABC Contractors

Figure 10-12

Contractor Bid Breakdown

ABC Contractors

Client Residence

The following is a breakdown of labor and material costs for the above mentioned project.

1.	Permits	$ 5,380
2.	Site Preparation	6,841
3.	Excavation	8,940
4.	Concrete	22,750
5.	Utilities	4,023
6.	Water and Sewer	18,211
7.	Rough Carpentry	108,550
8.	Windows	29,261
9.	Roofing	7,698
10.	Steel	3,426
11.	Fireplace	6,080
12.	Gutters	1,563
13.	HVAC	19,934
14.	Plumbing	27,741
15.	Electrical	26,537
16.	Garage Doors	2,586
17.	Insulation	5,391
18.	Drywall	14,773
19.	Trim Carpentry	18,187
20.	Cabinetry/Tops	22,511
21.	Flooring/Ceramic Tile	18,305
22.	Driveway	8,189
23.	Walk	1,500
24.	General Conditions	5,235
25.	Painting	15,441
26.	Shower Doors and Mirrors, Accessories	8,008
27.	Overhead and Profit	63,273

Total Cost of Project $ 480,334

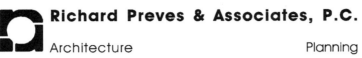

Richard Preves & Associates, P.C.

Architecture   Planning

Figure 10-13
Contractor's Bid Analysis
Client Residence Bid Tabulation and Analysis

Trade	ABC Contractors ($ 480,334) Original Bid $	Amended Bid $	Remarks	Done-Right Contractors ($ 432,458) Original Bid $	Amended Bid $	Remarks
Permits	5,380			1,830	+ 3,550	Too Low
Site Prep	6,841			1,200		
Excavation	8,940			7,900		
Concrete	22,750			30,300		
Utilities	4,023			1,200		
Water and Sewer	18,211			7,300		
Rough Carpentry	108,550			130,000		
Windows	29,261		Pella Windows	29,501	+ 4,417	Nordco Windows
Roofing	7,698			8,300		
Steel	3,426			9,400		
Fireplace	6,080			2,900		
Gutters	1,563			2,100		
HVAC	19,934		Includes Humidifier	15,630	+ 900	No Humidifier
Plumbing	27,741			20,500		
Electrical	26,537	+ 1,231	Light Fixtures Adjust	21,000	- 2,355	Credit
Garage Doors	2,586			2,800		
Insulation	5,391			4,950		
Drywall	14,773			16,000		
Trim Carpentry	18,187		Door Hardware Included	17,165	+ 4,725	Door Hardware Excluded
Cabinetry/Tops	22,511			15,900	+ 5,800	Not as Specified
Flooring/Ceramic Tile	18,305			16,690		
Driveway	8,189			6,400		
Walk	1,500			0		
General Conditions	5,235			2,325		
Painting	15,441			24,500		
Water Softener			Included in Plumbing	1,425		
Shower Drs, Mirrors, Access	8,008			3,108	+ 2,750	No Shower Doors
Overhead & Profit	63,273			32,134		Probably not Truly Stated
Bid as Submitted	480,334			432,458		
Adjusted Bids		+ 1,231			+ 19,787	
Contractor's Fee Adjust		0			+ 2,968	
Adjusted Total Bid		481,565			455,213	

tify who your daily contact will be throughout construction, and decide whether you can form a trustful relationship with this person. References should be checked with telephone conversations and better yet, in person at a completed project. Question his current and projected workload. Does he have the staff to adequately supervise the project or are they overextended? Are you confident that you will receive his undivided attention? The answers to these questions may make a difference when trying to decide between two finalists. Although one may be slightly more expensive, it may be worthwhile to pay the additional amount if you feel confident this contractor will be better to work with.

The next question can pay big dividends. Ask each contractor if he sees any opportunity to save money. In many instances, a contractor may know a less expensive method for building what you want. For example, a contractor may suggest a substitute for a material at a lower price. He may know that a light fixture produced locally is similar in appearance and far cheaper than the fixture specified. Or he may ask if you are aware that those bronze doorknobs the architect insisted on cost $350 each? After going through the design process, contractors can be a good reality check. Do you really want that hot pink whirlpool tub? Ask an experienced general to help you economize and he will usually have a few good suggestions.

Since construction prices always continue to grow more expensive, general contractor proposals can be negotiated, much like car prices. The larger the project, the more room for a downward adjustment. Let's say that a general has 7 percent profit built into a $300,000 project, equaling $21,000. On a project costing $500,000, that figures jumps to $35,000. If the level of finishes is driving up the cost, the contractor may be inclined to reduce his profit, because his time spent on the job is about the same whether the finishes are medium or expensively priced.

Especially if there is tight competition for your project, asking a contractor to reduce his profit margin can only net you positive results. Try saying something like this: "We have narrowed down our selection to you and XYZ Company, but you are slightly higher in price. Can you adjust your bid to get this project?" Anticipating these last minute adjustments, generals often build a small amount into their original bid just to be able to magnanimously offer a reduction. You have nothing to lose for trying!

Review their warranty program. After reading chapter 12, I am sure you will not sign a construction contract before checking this important document. Construction takes several months, but you must live with the results for many years to come. Differences among warranty programs may be enough to steer you away from certain contractors.

Finally, verify their financial health. Since construction involves large sums of money, it is very important that your contractor have the capital resources to meet labor and material bills between payments received from his clients. Newer construction companies frequently operate on a shoestring. One miscalculation on a bid can put them in serious financial trouble.

I once designed a project for a couple who bought an old two-flat in a historic area of Chicago for renovation into a single family home. They chose a casual friend to construct the project. His crew would work diligently until they received a payment from the homeowner, only to disappear for days at a time. Progress would lag until a month went by and another payment was due. Then the crew would return and work like crazy to justify another payment, only to disappear again. As months went by, the problem worsened. It finally reached the point where no work was occurring.

After I finally tracked the contractor down at another job, he admitted he had serious financial problems.

He was using funds from one project to pay the bills on another project. He lacked sufficient cash to survive between payments, so he was "robbing Peter to pay Paul." If you are unsure about the financial well-being of a contractor, request a statement indicating their current financial position. Avoid firms that have little cash on hand.

You have now made your final selection to complete your project team. Hopefully everyone is on the same page, united to produce the best project possible for you. You've finished most of your management work. Next up, construction!

Chapter Ten Recap

- **Builders and general contractors can determine project cost using three different methods. Each has both advantages and disadvantages to consider.**

  - *Lump Sum*
  The most common method of project bidding, this system determines a fixed cost based upon the architectural drawings. Multiple competitive bids can be obtained, but the guaranteed price is only as reliable as the drawings they are based upon.

  - *Cost Plus*
  This method can be used when the scope of the project is not completely defined at the start of construction. It is very useful when a project must be completed in a short period of time. The builder or contractor will pass all the construction costs directly to the homeowner, adding a percentage to the total for their overhead and profit. The total project cost is not determined until all decisions have been made, usually well into the construction phase. Therefore, the final cost is a moving target. Competitive bidding is not present in this method.

  - *Open Book*
  This method is an extension of Cost Plus, and is more commonly seen on commercial construction projects. The builder or contractor will solicit multiple subcontractor bids for every major trade and share these proposals with the homeowner. This is a good way to put some competition back into the Cost Plus option.

- **Construction pricing is based upon the accuracy and content of the architectural drawings. At a minimum, project drawings should include the following:**
  - **Site Plan**
  - **Schedules**
  - **Floor Plans**
  - **Details**
  - **Exterior Elevations**
  - **Sections**
  - **Specifications**

- **Construction bids must be analyzed to confirm they are on an "apples to apples" basis. Your professional can determine if the bids are comprehensively based on the project's scope by using cost breakdowns on a trade by trade basis furnished by each bidder.**

- **Before making a final selection, interview each general contractor candidate to verify their bids. Meet the key staff that will be working on your project everyday to develop a level of trust.**

- **Examine their financial condition to verify the construction company is solvent.**

Chapter 11

Ensure Construction Quality, Budget and Schedule without Picking Up a Hammer

Imagine the satisfaction of watching your dream project being constructed. After all the time you spent assembling a team and plodding through the design process, you can finally see concrete, lumber, and bricks assembled to form your new living environment. But homeowners have a tendency to let down their guard when construction starts. Their own actions frequently open the floodgates, destroying budgets and schedules. Maintaining your budget, schedule and quality is a key ingredient in our six-step process.

It has often been my experience that homeowners tightly control the budget during the design phase, constantly worrying about the final project cost. They make many hard decisions, carefully weighing each design option versus the additional cost. Then, as the project enters construction, they change course as if they suddenly had won the state lottery. The budget, so carefully monitored during design, is thrown out the window as changes are authorized practically every other week. The following is a story that perfectly fits this scenario.

I was hired to design a custom house on a beautiful site overlooking a lake. My client was a successful businessman who recently had sold his company and wanted to enjoy his retirement in a new home. He was a good client to work with because he applied the same methods for running his business to designing his dream home. Goals and objectives were clearly established along with an iron discipline not to exceed the budget. Every design issue was carefully discussed in relation to its impact on the final cost. No addition to the design was made without correspondingly deleting an item of equal value. I did not have to remind this couple about exceeding their budget. I already had a bulldog on the job!

General contractor bids came in right on the money, and construction began without a hitch. As the framing was nearing completion, my budget bulldog suddenly started to dig up hidden bones and authorize design changes. The contractor was kept busy estimating the cost of moving walls, adding windows, and enlarging rooms. When I pointed out to my client that these charges were overpriced, he didn't seem to care. I was baffled.

After the extra modifications reached nearly 20 percent of the original budget, my curiosity could no longer be kept in check. When I asked why he now allowed the budget to soar after watching every penny during design, he gave me a surprising answer: "I never thought that any project's construction bids came in on budget. Assuming that your estimates would be too low, I reserved 20 percent just to cover the original design program. When the bids came in at your estimate, I had money to burn!"

Some homeowners seem to play games with how much they actually have to spend on their project. If you hold back funds during design, either through indecision or intentionally to control the budget, you stand to lose money going on a spending spree for changes during construction. I can understand why homeowners want to build some cushioning into their budget, but they lose their advantage by paying greatly inflated prices to the contractor for those late changes. No matter how much I preach the importance of making design modifications during design, homeowners love to make changes throughout construction. Not only are they at the mercy of the builder's or contractor's fees for what they insist on changing, but they also delay their schedule by slowing construction.

Since I can't get anyone to follow my advice, I am going to provide you with some strategies to protect your pocketbook for those inevitable construction changes. When you are first interested in deviating from the original design, I recommend you adhere to the following procedure.

First, consult with your design and construction professionals, requesting their advice on the merits of the modification. As an architect, I am always surprised to learn about changes from the contractor, not my client. I think they are often embarrassed to ask me to alter my design, preferring to deal directly with the contractor. Often, I don't discover the change until I make my next site visit, only to find the modification already built. This can lead to one of many problems caused by hasty design alterations. Moving walls, doors, and window openings is not always as simple as it seems.

Structural issues affecting wall placements and openings must have properly designed structural supports. Unless the contractor anticipates these problems,

moving framing can have serious consequences. Changing dimensions of rooms can also cause other problems, for example, expanding a kitchen means the cabinetry has to be redesigned and plumbing and electrical outlets may have to move. On a construction project, one seemingly simple change can affect the work of several tradesmen in ways you never anticipated. It is important to revise the drawings before the change is made at the construction site.

The second step I recommend if you want to make changes during construction is to ask the contractor to prepare a *change order*. Figure 11-1 shows this comprehensive, written summary of a proposed modification. You will notice it includes information about four critical subjects. First, a description of the proposed modification is listed. Second, the cost of the change is itemized. Third, the cost is reflected in a newly adjusted total project cost. The fourth item shows the impact of the change on project timing. If the schedule must be extended to accommodate the change, the actual delay is stated in working days. Finally, your signature at the bottom of the form authorizes the contractor to proceed with the modification.

Homeowners run into trouble when the change is made before this form is completed. Often, revisions are requested by the homeowner at critical times, and some contractors, being notorious for their slow turnaround time in producing paperwork, will delay determining the cost of the modification until after it is completed. Thus, you have created an obligation to pay for the change, regardless of the cost, when the bill is presented. This would be similar to ordering a deluxe dinner at a four-star restaurant without checking the prices on the menu! Requiring a written change order from the contractor prior to the alteration may slow construction for a day or two, but will pay off in cost savings. Avoid verbal "ballpark" numbers. These are not official.

With a predetermined price, you can either accept or try to modify the cost of the proposed change order.

Figure 11-1

Change Order Authorization for Construction Revision

ABC Contractors

Number: 1
Date: 10/7/00

Project Name: Client Residence

Contractor: ABC Contractors

Date of Contract: June 26, 2000

The following revision to the construction contract has been authorized:

- Add dividers and baskets and adjust shelving layout per attached decisions and discussions.

Original contract amount:	$ 453
Net change from previous authorized revisions:	$ 0
Adjusted contract amount before this authorized revision:	$ 453
Net change for this authorized revision:	$ 0
New contract amount:	$ 470,099

This revision will change the time of construction by: 2   days

The undersigned agree and approve the terms of the above authorization for construction revision.

_____   _____   _____

Owner   Contractor   Architect

Signature: _____   Signature: _____   Signature: _____

Date: _____   Date: _____   Date: _____

Once you see the actual cost, you may change your mind. General contractors and their subcontractors usually do not give fair value for change orders. Since they are not in a competitive situation and revisions frequently interfere with their future work schedule, get ready to be overcharged from 20 percent to 40 percent, depending on the size of the modification. As the contractor sees it, if you want to pay the freight, that's fine. If you choose not to accept his inflated prices, that's fine also. His next job awaits. Change orders are often submitted on a take-it-or-leave-it basis. Attempts to negotiate change orders usually run into a brick wall!

Let me warn you about one other facet of change orders. Once on a remodeling project, I had an irate call from my client regarding an unexpected change order he had just received from the contractor. He was being billed for a new exhaust fan in his master bathroom even though the plans called for the original to remain. Tracing the story, I discovered that the electrician working in the bathroom made a casual comment to my client that the old fan was looking a little worn and replacement might be a good idea. By replying, "Sounds good to me," the homeowner had unintentionally created a change order.

The electrician thought he had been authorized by the homeowner to replace the fan at additional expense. My client assumed the electrician was going to throw in the new fan for free! Neither party even thought to ask about money, and since the replacement was installed the next day, no paperwork was every initiated. Beware of idle remarks to tradesmen on the site. One casual comment can become a costly surprise.

Besides expanding your budget, change orders ruin your project's schedule. Since existing construction must be modified or extra construction added, your schedule will lose momentum. Before beginning a project, a builder or general contractor will inform each subcontractor of an approximate date he will be needed at the site. Although subs do not always appear exactly on the appointed day, they do try to keep their schedule prioritized to the projects they receive. If one trade has to extend their work on the project due to change orders, all subcontractors who planned to appear on a certain date must now adjust their schedules.

For example, let's assume that a homeowner requests revisions that require the carpentry subcontractor to modify walls that are already erected. This change causes the carpenter to remain on the job several days longer than originally anticipated. The electrician, plumber, and heating trades who were scheduled to follow the carpenter now must wait because they can't start. In a busy economy, these trades do not sit around waiting for the modifications to be accomplished. They go to the next job or try to pick up another project they can work on now. When the carpenter finishes his additional work, your job site will be vacant for a while, waiting for new subs to appear. This process is cumulative, as it affects all trades downstream from the change. Multiple change orders affecting several trades will make matters even worse.

Nothing kills a schedule like change orders. Since the majority of these revisions are initiated by the homeowners, delays are usually their own fault. If you have ever watched developers who build tract houses in large subdivisions, you know they can complete a new house, start to finish, in four to five months. Custom projects can take at least twice as long. One big reason for the difference is that the developer does not have to cope with homeowners issuing changes. Since modifications seem to be inevitable on custom projects, don't complain about the length of construction. It could be your fault!

Monitoring Workmanship

The state of construction workmanship in today's market is always a hot topic for discussion. We could talk at great length about the effects of trade specialization, or the demise of trained union labor versus less skilled tradesmen. Part of the debate could focus on today's superior tools and better-engineered materials. We have previously considered the old saw that "you get what you pay for." Regardless of how you stand on these issues, you probably wonder how average homeowners can tell if quality construction is being provided on their project.

Many homeowners embarking on a building project assume that the quality of construction will be monitored by the local building department during its inspections. This assumption is only partially true. Governmental building departments are established on the concept of protecting public safety. My definition of their scope of review falls into the following three categories:

- Structural design and construction of foundations and framing to prevent collapse or unstable conditions.
- Health and sanitation through safe plumbing practices to provide clean drinking water and the removal of waste.
- Fire and health safety furnished by properly designed and installed electrical and heating systems.

With the strict regulations of today's building codes and the training received by many building department inspectors, these three categories are usually well controlled. After all, how often do you hear about a house that has collapsed? Building department inspections occur at several specific construction milestones, summarized as follows:

- Excavation prior to placing foundations.
- Inspection prior to pushing dirt against the exterior foundation walls, known as *backfilling*.
- Rough carpentry framing and rough-in inspection of electrical, plumbing and heating work.
- Insulation inspection before drywall.
- Final occupancy approval.

This sounds like a comprehensive list, but actually there are a lot of areas not covered. You may have noticed that the list does not include "cosmetic" or finish elements, such as how straight the walls look, the quality of drywall work or the tightness of wood trim joints. These items do not fall within the realm of public safety. Unless you have a cousin who is a tradesman, you can choose between assessing workmanship yourself or hiring someone to do the job for you. A good rule of thumb is that the building department focuses on construction that is largely hidden from view. The architectural finish workmanship you see everyday is your responsibility.

Consider one final word on building department inspections. Even though inspectors have greatly improved over the years, not all are created equal. Contractors and subcontractors all love to make friends with inspectors, hoping to receive easier treatment at inspection time. Or if your inspector attended a late party the night before your inspection, he may not be as sharp as he should be. I have seen the same inspections on similar model homes in the same subdivision take either five minutes or fifty minutes to accomplish. Sometimes attention to detail can be arbitrary. Building departments are much like school systems in your neighborhood, some are better than others.

If you feel confident after the qualification and interviewing process that your builder or general contractor is known for good workmanship, you probably will not have to require independent, extensive inspections. But if you are uncomfortable with your construction professional, you can hire an on-site representative to check the quality of workmanship.

The first option is to have your architect monitor construction. You will recall that within the Traditional System, a full-service architect can offer construction administration to you. The Design/Build System architect unfortunately does not, as the architect works for the builder and would not be welcome at the site as your representative. During our discussion on contracts in chapter 5, we described the architect's role as observing the contractor's compliance with the requirements and then reporting to you. Your architect does not control or supervise construction, but functions as an advisor. Since this service will require additional architectural fees, you must decide how often the architect will visit the site. Intervals could be tied to construction activity, as trades arrive on the site. This could be packaged into five to ten visits for the duration of the project. Weekly visits are best if you can afford them. Depending on the schedule, this requires fifteen to thirty trips, and the expense would be correspondingly higher.

Every site visit made by the architect should result in a written report. Figure 11-2 illustrates a *field report* form used by my office. The report first describes ongoing work to inform the homeowner how his project is progressing. A section titled "observations" contains information the architect felt needed attention or correction by the contractor. Items requiring further explanation or design attention are also listed. The final category, "information or action required," indicates each project team member's immediate responsibility. Upon completion, this form is distributed to the homeowner, contractor, and any other building professional that should be in the loop. By the completion of the project, these field reports provide a good abbreviated history of construction.

Problems relating to workmanship can be difficult to remedy. Construction thought to be unacceptable by the architect and requiring replacement could be met with resistance from the contractor. They will tell you that their work falls within the "standard of industry tolerances." These disagreements are often resolved on your part by what you can accept living with everyday. If you don't mind seeing a few imperfections in the drywall, the contractor prevails. The contractor loses, however, if you and your architect are perfectionists. There are no tried and true methods I can recommend to address this issue. Each case and the personalities involved are different. Use your common sense to reach a fair solution.

If you are using the Design/Build System, consider hiring an independent consultant to visit the site periodically. An architect or home inspector would fill the bill. Since he is a neutral party having no affiliation with preparing the drawings or selecting the building professionals, he is in a good position to give you unbiased advice. Although architects normally do not make a practice of observing projects designed by others, home inspection companies that prepare reports for residential real estate purchases may be acceptable for this role.

They would function in the same manner as if your own architect was on site. He can prepare reports based on his observations during site visits. Home inspectors usually charge on a pay-as-you go basis, although you could suggest a total fee tied to a specified number of site visits. You can find home inspectors in the yellow pages or by talking to local Realtors. Qualify their credentials just as you would check out any building professional.

The third option for controlling quality of workmanship relates to contractor or builder payment. Whether you have on-site professional assistance or you are representing yourself, you should pay only for acceptable workmanship. When your construction professional submits an application for payment, a topic we will discuss in the next section, you have the best opportunity to insist on corrections for faulty work.

For example, let's assume you have received a monthly invoice to cover construction that includes the installation of drywall and finishing trim, such as door and window moldings. During a site inspection, you or your representative notice that the corner joints of the door and window trim are not nice and tight as they should be. Instead of paying the entire amount billed, hoping that the builder/contractor will soon fix this problem, pay only for the portion of the work that is acceptable.

If the payment is $25,000, apportioned as $15,000 for drywall and $10,000 for carpentry trim, tell your

Richard Preves & Associates, P.C.

Architecture   Planning

Figure 11–2
Architect's Field Report

Project: Client Residence   **Distribution:**   **Owner**   **X**
                                                                          Contractor   **X**
Project No: 987654   **Consultant**

Field Report No: 18   **Date of Report:**   October 13, 2000

Work in Progress:

- Plumbing rough-in 80% complete.
- Electrical rough-in 45% complete.
- HVAC rough-in has just started.
- Siding installation started.

Observations:

- Window and skylight installation is complete.
- Fascia and soffit vent installation is complete.

Items to Verify:

- Location of light type "U" on the upper level. Install as low as possible below the triangular window.
- The wall at the bath tub and shower unit in Bath 108 will be furred out ± 2" at the toilet wall side.

Information or Action Required:

- General contractor should verify the size of siding nails.

Report by:

Richard Preves, A.I.A.

professional you are holding back $2,500 of the trim amount until the corrections are completed. This shows you are willing to pay for all of the drywall and part of the trim materials, but not for the poor trim workmanship. Don't expect this decision to be greeted with open arms by the builder/contractor. He will probably insist on full payment. However, it has been my experience that a partial payment is more acceptable than no payment whatsoever. Builders/contractors can more readily understand withholding funds to cover disputed work than a blatant refusal to pay at all. Unfortunately, homeowners sometimes mistakenly take this concept one step too far by trying to withhold the entire amount. This is unfair to the subcontractors whose work requested in the payment application is acceptable. Going to this extreme is like declaring war, and I do not recommend it! Contractors and subcontractors have a method of combating this strategy called liens, a subject discussed later in the chapter. Unless the workmanship on the entire project is a disaster, expect to pay for acceptable craftsmanship on time.

The Payout Process

At the various intervals predetermined by the construction contract, you will be presented with payment requests for completed construction. This process involves more than simply writing a check. Since the bulk of construction projects are funded by financial institutions, procedures have been established to protect their investment. Even if you are acting as your own lender and paying cash for your project, you will want to benefit from the same safeguards.

Included within figure 11-3 is the step-by-step procedure for the application and funding of a pay request. Several forms must be properly completed by the construction professionals and their subcontractors before money comes their way. You will be startled by the amount of paper required to complete this process. Builders and general contractors who are paid on a monthly schedule usually start working on the next payout before the last one is funded.

The first step finds the builder/contractor submitting a written request for payment. These written demands are created in many different formats. If a bank or title company is in the picture, standardized forms are available to facilitate the process. Figure 11-4 and figure 11-5 comprise the most commonly used documents for this purpose. *Application and Certificate for Payment* depicted in figure 11-4 works in combination with the *Contractor's Sworn Statement to Owner*, shown in figure 11-5.

The application and certificate form indicates basic information for the project. All parties involved in the project are listed, as well as the dates covering the work period. A table of figures at the right includes the total project cost, change orders, previous payments, the amount requested and the balance due. You can tell at a glance the exact financial history of the project. An area in the left-hand portion of the page is available to itemize change orders that have been approved. This particular form is designed for use by an architect certifying the payment request. Specific language for both the builder/contractor and the architect is included, accompanied by spaces for signatures and notary seals. Forms are also available for the owner acting as his own authorizing agent.

The contractor's sworn statement in figure 11-5 is a comprehensive summary of the amounts for every trade and subcontractor. For instance, the carpentry trade lists the name of the subcontractor, the total amount of their respective contract, past payments, requested amount, and the balance to finish. This form is invaluable as it enables the reviewer to determine whether the values requested for each trade correspond to the amount of work completed at the site. If the carpentry work accurately represents 80 percent of the completed work as requested, the amount is approved. When the plumber's request of 85 percent is compared to only 65 percent completion in the field, a downward adjustment is required.

It is common to ask for corrections in the sworn statement prior to approval. Frequently, due to the length

Figure 11-3

General Contractor Payout Process

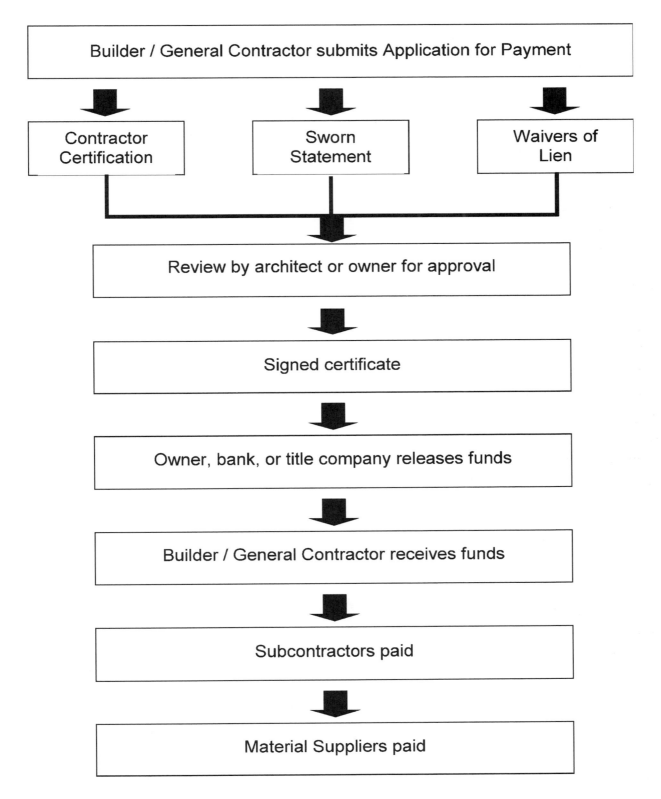

Figure 11-4
Application and Certificate for Payment

TO (Owner): Joe & Mary Client	APPLICATION NO.: 4
	PERIOD TO: 10/31/00
CONTRACTOR: ABC Contractors Inc.	CONTRACT DATE: June 26, 2000
ARCHITECT: Richard Preves & Assoc.	

CONTRACT FOR:

Application is made for Payment, shown below, in connection with the contract. Contractor's sworn statement to Owner is attached.

CHANGE ORDER SUMMARY:

	ADDITIONS	DEDUCTIONS
Change Orders approved in previous months by Owner		
TOTAL_____		

Approved this month

Number	Date Approved		
1	10/9/00	850	
Total Net Change Orders		850	

1. Original Contract Sum	468,796
2. Net Change by Change Orders	850
3. Contract Sum to Date (Line 1+2)	469,646
4. Total Completed & Stored to Date	140,894
5. Less Previous Certificates for Payment (Line 4 from prior Certificate)	70,430
6. Current Payment Due	70,464
7. Balance to Finish	328,752

The undersigned Contractor certifies that, to the best of his knowledge and belief, the information and work covered by this Application for Payment has been completed in accordance with the Contract Documents, that all amounts have been paid by him for work for which previous Certificates for Payment were issued and payments received from the Owner, and that current payment shown is now due.

CONTRACTOR:

By: ABC Contractors   Date:_____

State of Illinois, County of Lake
Subscribed and sworn to before me this 2rd day of November

Notary Public:

My Commission Expires: December 15, 2002

ARCHITECT'S CERTIFICATE FOR PAYMENT
In accordance with Contract Documents, based on on-site observations, the Architect certifies to the Owner that to the best of Architect's knowledge, information and belief, the work has progressed as indicated, the quality of work is in accordance with the Contract Documents, and the Contractor is entitled to payment of the CERTIFIED AMOUNT.

AMOUNT CERTIFIED: $70,464

ARCHITECT:_____ DATE:_____
The Amount Certified is payable only to the Contractor named herein.

Figure 11-5
Contractors Sworn Statement to Owner

Project: Client Residence

Payment Application No: 1

Page 1 of 1

8/1/2000 thru 8/31/2000

Description of Work	Subcontractor	Original Contract Amount	Authorized Revisions	Adjusted Contract	Completed Amount	%	Past Payments	Amount Requested This Application	Balance to Finish
Permits		$ 2,188.00					$ 2,188.00	n/a	n/a
Excavating	Deep Excavating	$ 8,994.00					$ 4,000.00	n/a	$ 4,994.00
Site Utilities		$ 2,700.00					n/a	n/a	$ 2,700.00
Concrete	Durable Construction	$ 28,735.00					$ 7,572.00	$ 10,823.00	$ 10,340.00
Electric	Lake Electric	$ 27,735.00					$ 2,000.00	$ 6,000.00	$ 19,735.00
Plumbing	Pipes To You	$ 29,634.00					$ 13,685.00	$ 1,500.00	$ 14,449.00
HVAC	Reds Heating	$ 21,232.00					$ 9,500.00	$ 5,000.00	$ 6,732.00
Drywall	JM Construction	$ 12,745.00					$ 6,000.00	$ 4,000.00	$ 2,745.00
Roofing	Tops Roofing	$ 8,115.00					$ 4,900.00	$ 3,215.00	n/a
FirePlace	Warm Glow	$ 10,539.00					n/a	$ 10,539.00	n/a
Cabinets	R & Q Cabinets	$ 39,393.00					$ 6,000.00	$ 18,393.00	$ 15,000.00
Windows	Glass, Inc.	$ 29,631.00					n/a	n/a	$ 29,631.00
Carpentry	JM Construction	$ 101,602.00					$ 7,500.00	$ 6,620.00	$ 87,483.00
Steel	Strong Stuff, Inc	$ 2,581.00					n/a	n/a	$ 2,581.00
Insulation	Smith Insulation	$ 4,957.00					$ 2,085.00	n/a	$ 2,872.00
Appliances	XYZ, Inc.	$ 9,200.00					n/a	n/a	$ 9,200.00
Finish Hardware	Knobs Are Us	$ 8,262.00					n/a	n/a	$ 8,262.00
Flooring and Tile	Floors, etc	$ 17,728.00					n/a	n/a	$ 17,728.00
Waste Removal/Cleaning	We Pick 'Em Up	$ 2,800.00					n/a	n/a	$ 2,800.00
Driveway	Smooth Drive Co.	$ 7,500.00					n/a	n/a	$ 7,500.00
Supervision Overhead and Profit	JM Construction	$ 93,374.00					$ 5,000.00	$ 4,374.00	$ 84,000.00
Total		$ 469,645.00					$ 70,430.00	$ 70,464.00	$ 328,752.00

n/a = non-applicable

153

of time required to create this paperwork, subcontractors will apply for an amount they anticipate will be installed by the time the application is actually reviewed. Sometimes they guess wrong by over estimating their future work program. Just as often, subcontractors will be too late supplying their paperwork to their builder/contractor and miss the payout altogether. This is reflected by a value on the application below the amount of work completed. This scenario works to your benefit. All subcontractors know that no paperwork equals no pay! Occasionally construction professionals will artificially inflate their requested amounts, trying to get more money in advance. Careful analysis of the sworn statement in comparison to completed work at the site will enable you or your representative to spot this problem.

Waivers of Lien

The third component of payout paperwork is the *waiver of lien*, also know as a mechanic's lien. Construction liens and lien laws are complex subjects; in fact, entire books and seminars are dedicated to the topic. Let's reduce this concept to a simple explanation. Everyone who either supplies materials or labor on your project is entitled to be paid. A lien is a legal method provided by state law that entitles unpaid companies or individuals to file a document placing a "cloud" on your real estate property title recorded at the county courthouse.

For example, let's say the lumberyard has not received payment from the carpentry subcontractor or builder/contractor. After a certain period of time, (typically ninety days after last furnishing materials), the lumberyard can file a lien against your property. Another example that occurred on one of my projects involved a plumbing subcontractor who was dissatisfied with the work of a newly hired tradesman. He fired the tradesman and refused to pay his wages for the week he was employed on my project. The disgruntled, unemployed plumber slapped a lien on my client's title. This did not go over particularly

well with my client! A third example could involve an owner who refused to pay the contractor. In this case, not only would the contractor file a lien against the owner's property, but every affected subcontractor and material supplier would also file a separate lien.

Now why should your title be affected by someone else's negligence? The reasons and remedies are too complex a subject for the confines of this book. What I want you to take away from this discussion is the fact that unless payments are carefully administered, these events could occur on your project. For your protection, a form called a *waiver of lien* is commonly used during the payout process. Figure 11-6 is an example of two waivers of lien forms combined into one. A *Partial Waiver of Lien* is used for progress payment to contractors and subcontractors as they continue working on the project. A *Final Waiver of Lien* is submitted by each company for its last payment on the project. The form lists pertinent project information, indicating the name of each company, the type of work it performed, and the dollar value for this particular payment.

As you read the fine print, you will see that each company acknowledges that you have paid a certain amount and it releases you from any liens or claims associated with this portion of the project. As figure 11-6 shows, this example waiver is from the electrician for an amount of $16,000. The subcontractor warrants that the money you paid has been used to pay for all labor and materials associated with his work. Thus, you are protected in case a subcontractor does not actually pay what he owes to his own employees and material suppliers. For every trade listed on the contractor's sworn statement, a waiver of lien should be furnished in the exact amount requested.

Does this sound like a lot of work? For contractors, it is just one more chore to complete in order to get paid. For the homeowner, these liens are usually examined by the title company and used by financial

Figure 11-6

Partial or Final Waiver of Lien

STATE OF
COUNTY OF

TO WHOM IT MAY CONCERN:

Whereas the undersigned has been employed by **ABC Contractors, Inc.** to furnish **Electrical Construction** for the premises know as **1234 Main St., Somewhere, US 12345** of which **Mr. Joe Client** is the owner.

The undersigned, for and consideration of **Six Thousand ($6,000.00)** Dollars and other good and valuable consideration, the receipt whereof is hereby acknowledged, do(es) hereby waive and release any and all lien or claim of, or right to, lien, under the statutes of the State of **Illinois**, relating to mechanics' lien, with respect to and on said above-described premises, and the improvements thereon, and on the material, fixtures, apparatus or machinery furnished, and on the moneys, funds or other considerations due or to become due from the owner, on account of labor, services, material, fixtures, apparatus heretofore furnished, or which may be furnished at any time hereafter, by the undersigned for the above-described premises.

Given under my hand and seal this **2nd** day of **November, 2000**.

Signature and Seal: _____

Contractor's Affidavit

STATE OF
COUNTY OF

TO WHOM IT MAY CONCERN:

The undersigned, being duly sworn, deposes and says that he is **President** of **Lake Electric** who is the contractor for the **Electrical** work on the building located at **1234 Main St., Somewhere, US 12345** owned by **Mr. Joe Client**. That the total amount of the contract including extras is **$27,735.00** of which he has received payment of **$2,000.00** prior to this payment. That all waivers are true, correct and genuine and delivered unconditionally and that there is no claim either legal or equitable to defeat the validity of said waivers. That there are no other contracts for said work outstanding, and that there is nothing due or to become due to any person for material, labor or other work of any kind done or to be done upon or in connection with said work other than above stated.

Signed this **2nd** day of **November, 2000**.
Signature: _____

Subscribed and sworn to before me this **2nd** day of **November, 2000**.

_____
Notary Public

institutions to disperse funds. Although you must pay for this service, lenders do not give you the choice. You must use a title company. As each application for payment is submitted by the builder or contractor, the title company checks the courthouse records to verify that no liens have been filed against your property. In the event a lien has been filed, the payout will not be funded until the lien has been removed and the claim satisfied. With a waiver of lien in hand, you are protected. Without the waiver, a claim stands against your real estate title until it is released.

Title companies are very useful for the homeowner and the lender because they do the majority of the work associated with a payment request. They also check all the figures on the application and sworn statement to verify the math is correct. If you have the choice not to employ a title company, you are now responsible for reviewing the accuracy of each payment's pile of paperwork. Title companies charge by the size of the project and the number of payouts. Each payout process will cost several hundred dollars, so the fees for a large project with many payment applications will start to add up.

Once all three paperwork components of the payout have been submitted, reviewed, and finally approved by you (and your representative, if you use one), the funds are paid to the contractor or builder. If you are using a lender, the title company will issue checks to each applicant from funds furnished by the financial institution. In case you are paying directly for construction, you write the check. Once the contractor or builder and their subcontractors receive their money, they in turn pay their material suppliers.

By now, I hope you appreciate the importance of properly administrating payments to your construction professional. On smaller projects, once you have been through the drill of the first payout, subsequent payments are easier. For large projects in the Traditional System, enlisting the help of your architect will ease your burden. Unless you recruit an independent advisor when using the Design/Build System, you are on your own in reviewing completed values on the sworn statement. In either event, make sure all the *i's* are dotted and the *t's* are crossed before you authorize the release of your money.

Controlling Time

So far, we have discussed cost and quality issues affecting your project. The remaining concern is the issue of time. This chapter should really be titled "Avoid the Cliché of Projects Costing Twice as Much and Taking Twice as Long." Time on construction projects is the most difficult variable to control. Unfortunately, as much as you may want to, you cannot hold a gun to the heads of your team and force them to work. However, you can continually monitor the progress and also make sure your actions do not cause delays.

The best strategy is to make time constraints as important as design and budget issues. Each of these components should receive equal attention from your professionals. As we have discussed at many points along our journey through the project process, many decisions have consequences that affect time schedules. It is best to know about these elements from the start and plan ahead for their prompt resolution.

With the help of your project team, determine a realistic timetable for the entire process from start to finish. Figure 11-7 demonstrates how a project schedule is assembled. The timetable begins with preliminary design, progresses through completion of contract documents and bidding within the Traditional System and estimates construction time. As your project evolves, actual progress can be compared against the original assumptions. This schedule can be continually updated as you either gain or lose time along the way. If the design phase took an extra month, you know that the completion of your project will occur one month later.

Once the start of construction is anticipated, a more comprehensive schedule can be furnished by the

Richard Preves & Associates, P.C.

Architecture   Planning

Figure 11-7
Project Timetable

Client Residence

1.	Preliminary Design	January 17, 2000
2.	Design Development	February 7, 2000
3.	Begin Contract Documents	February 21, 2000
4.	Contract Documents 50% Review	March 14, 2000
5.	Final Review of Documents	April 4, 2000
6.	Out for Bids	April 11, 2000
7.	Bids Due	May 1, 2000
8.	Award Contract for Construction	May 10, 2000
9.	Permits Received	May 24, 2000
10.	Begin Construction	May 27, 2000
11.	Complete Construction	November 25, 2000

builder or contractor. Figure 11-8 shows a *Construction Progress Schedule*, which is a detailed timetable. Each subcontractor trade is listed in a column along the left-hand side along with time expressed in weeks across a row on the top. You can see the time when the builder or contractor estimates each sub will appear during construction and the duration of his work. Some trades, such as carpentry, electrical, plumbing, and heating are scheduled for multiple appearances throughout construction.

Two time symbols are displayed on the schedule. The solid black blocks represent the contractor's estimate of the construction schedule. This shows the anticipated appearance date of each trade, established before construction starts, and assumes no delays due to material shortages, unusual weather, or owner requested change orders. Using the chart, you can predict construction milestone events. For example, the roof should be completed by the tenth week, and drywall installation should begin in the fifteenth week. Lightly shaded lines represent the actual date that each task was performed. You can see from the schedule in figure 11-8 that construction began on time but the progress is at least one week behind the original timetable.

The construction progress schedule enables you to monitor the course of activities on a weekly basis. Typically, delays in the schedule at one point are frequently made up down the road. Just as often, delays can pile up and compound the tardiness. As construction progresses, you can continually update the schedule right along with the builder or contractor. No matter the size of your project, insist that your construction professional prepares this invaluable document. Besides setting a benchmark for completing the project, it also commits the professional to carefully planning the weekly progress in advance.

Time delays in most projects occur during construction for the various reasons we have previously discussed. If actual construction events fall seriously behind the original schedule, call a meeting with your builder or contractor. Discuss what issues are causing the project schedule to be delayed. Two culprits are usually to blame. The first is controlled by the construction professional, the second by you. If you recall our discussion from chapter 4 regarding the contractor's duties, one of his main challenges is having the subcontractors show up on time. But no matter how much a contractor may try, he is at the mercy of his subcontractors' schedules.

A typical scenario between owner and contractor at such a meeting may go something like this: "Why is the project four weeks behind schedule?" the homeowner asks. The contractor replies, "The carpenter was two weeks late arriving on the job, plus the plumber and drywaller were both a week late. There's not much I can do." Unfortunately, this statement is all too true. The only method of counteracting this project phenomenon is to hire builders or general contractors who utilize many of the same subs on many projects. This gives the construction professional more leverage, because the sub's schedule really becomes his to control. During your professionals' qualification and interviewing process, ask each prospective builder or general contractor how well he can control his subcontractors' schedules.

The second cause of project delays is a subject we have already discussed—the homeowner's request for changes after construction has started. If you insist on making several design revisions during construction, don't expect to hold the builder or contractor accountable for maintaining the schedule. Change orders ruin the momentum of construction, delaying the schedules of the affected subcontractors. Typically, once the original construction schedule runs off the track, building professionals are far less motivated to sustain a timetable. The prevailing attitude is: "Your changes ruined my schedule, so you can live with the consequences." If you truly want to be in control of your project, control yourself!

Watching your project take shape during construction can be a very satisfying experience for you and

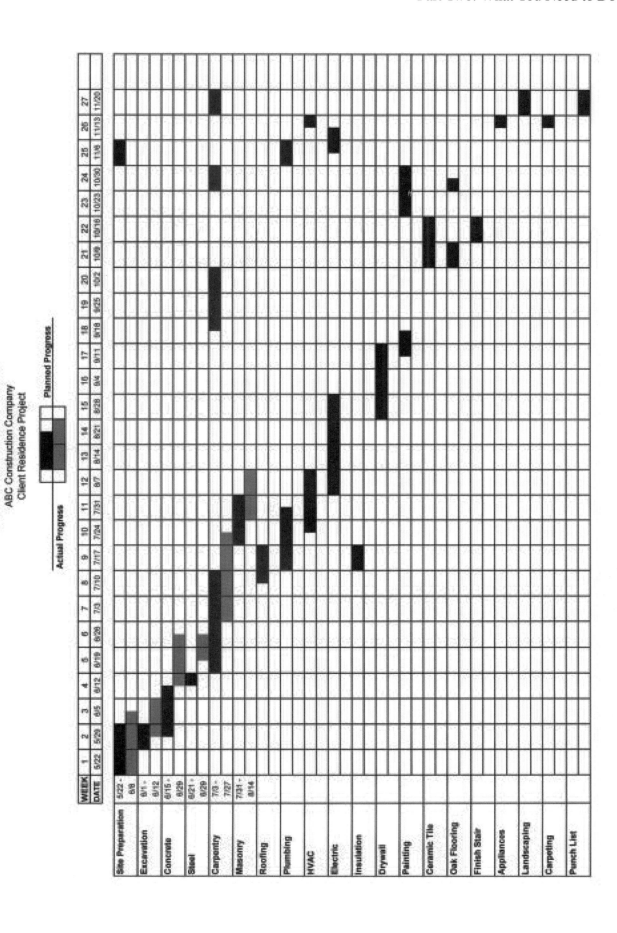

Figure 11-8
Construction Progress Schedule
ABC Construction Company
Client Residence Project

your project team. A well-organized, efficiently managed project takes on its own personality, as does one that is less controlled. I have given you all the tips to make yours fall in the first project category. Now it's up to you to use them!

Chapter Eleven Recap

- **Changes in the design during construction are common and should be reflected in a written document called a *Change Order*. This authorizes the builder or general contractor to alter the project cost and time schedule due to the proposed revision. Always have the proposed revision priced out in writing before approving the change order.**

- **Local building department inspections only cover a portion of construction workmanship. This includes structural, sanitation for plumbing and the safety of electrical systems. "Cosmetic" workmanship of finish materials such as trim are not inspected.**

- **Consider hiring outside help to verify construction workmanship if you do not have your own architect.**

- **Pay the builder or general contractor for only acceptable workmanship.**

- **Payment for construction uses a pre-established sequence for reviewing and approving the paperwork submitted by the construction professional. Lenders frequently use title companies to review all documentation prior to releasing funds.**

- **Waivers of lien are an important component of the payment request process and protect your investment in the project.**

- **Estimated time schedules are frequently exceeded on construction projects. Make time constraints as high a priority as design and budget for your project. Insist on updated project schedules from your professional and closely monitor the progress.**

- **Homeowner-generated change orders frequently upset construction schedules and momentum.**

Chapter 12

Successfully Close Out Your Project

As in any other business, fellow builders and contractors occasionally meet over a beer to commiserate. Often the conversation turns to the problems homeowners cause after construction has been completed. "They call me nearly everyday with complaints about a scratch here, or an imperfection in the drywall there. It's enough to drive me crazy, and then they expect me to have a carpenter at their disposal whenever they call. No way that's going to happen!" one builder remarked. Another contractor said, "Well, it really comes down to your warranty. I had my program written to exclude problems like that. I just tell them to read the fine print and leave me alone!"

We have finally arrived at the long awaited finishing line where you accept the project and take occupancy. Hopefully, all those long meetings, debates over space versus budget, and hours spent monitoring construction have paid off. But we have one more step to take before we complete the process. Our goal is to make sure all construction required by the drawings has been completed by the time you move in, or is at least scheduled to be accomplished soon after. Once the last tradesman walks out the door, the warranty program you accepted from your construction professional is in force. Regardless of how the warranty is written, I can recommend a few techniques to make sure you receive the maximum coverage allowed under the program. If your warranty is very restricted or limited, we will discuss possible options to give you some measure of protection.

Wrapping Up Construction

Completing construction can occasionally drag on, taking forever to finish what seem to be just a few nagging items. To facilitate this process, a document called a *punch list* should be prepared. This is a comprehensive list compiled near the end of construction. By examining our example in figure 12-1, you will see each room listed, with construction that should be completed or repaired. This applies to the entire project, both inside and out, top to bottom. You will notice our punch list mentions some items that have not been installed, such as the powder room faucet, mirror, electrical switch, and outlet plates. The majority of the entries concern minor repairs and touch-ups for painting, caulking, and adjustment.

If you have been working directly with a builder or contractor and have been flying solo throughout construction, schedule a walk-through to compile a joint list. Architects can join the party and add their input if you have their representation during construction. The more eyes you have, the more complete the punch list will likely be.

Richard Preves & Associates, P.C.

Architecture   Planning

Figure 12–1
Punch List

Client Residence

February 5, 2000

General
1. Turn over to owner all operating instructions and manuals for all equipment, warranty cards and information.
2. Install all screens on windows.
3. Caulk around window frames.
4. Fill in recessed stone sill area under south entry window.
5. Repair and repaint gypsum board areas displaying taping marks, air bubbles, nail head shadows, voids, cracking, and "blobs" to areas circled or noted during walk-through, and others by additional inspection by the contractor.
6. Adjust all roller catches on doors for snug, proper fit.
7. Fully seal all door hinge pins.
8. Where HVAC louvers do not lay flush against wall, fill gaps with caulk.
9. Remove all window stickers and tape.

First Floor

Hall Bath
1. Re-install sandblasted glass in window.
2. Install new shower door.
3. Furnish missing plastic shelf to towel bar (two were specified on October 8, 2000 letter – schedule).

West Bedroom
1. Repair nail pop in gypsum board in closet, south wall near ceiling. Touch-up paint.
2. Paint top of south window sash covered by top track.
3. Touch-up paint on baseboard.

Hall
1. Cove light dimmer switch sticks – adjust for smooth sliding.
2. Repair recessed down light in front of mechanical closet door – light bulbs instantly burn out.
3. Repair crack in baseboard, north wall at corner by mechanical room door.

Powder Room
1. Install faucet, drain and sink.
2. Install mirror.
3. Relocate towel hooks to wall per owner's direction. Patch and paint door.

Closet
1. Reinstall electrical jamb switch.

Dining Room
1. North recessed down light – lamp burned out.
2. Adjust down lights for full 90° downward orientation.

Living Room
1. Repair full-length gypsum board defects at ceiling/east wall intersection.

Client Residence
Punch List – February 5, 2000
Page 2

Family Room
1. Caulk intersection of fireplace wall stone with ceiling.
2. Re-configure down light switching to match previous circuiting.
3. Touch up paint wall shelves, base and trim scratches.

Second Floor

Stairs
1. Fill all nail heads holes, dimple and gouge marks on top railing cap on stairs, and repaint as necessary.
2. Remove excess built-up paint from underside of railing cap and in reveals.
3. Sand underside of railing walls and repaint as necessary.

Hall
1. Center closet, cut down shelves at access door.
2. Adjust all roller catches on doors.
3. Repair tape joint over closet door.

Bath
1. Caulk holes under window frame.
2. Door lock not working properly.
3. Adjust pocket door, rubbing in frame.

Bedroom
1. Install missing switch plate.
2. North closet door rubbing on carpet.
3. Clean paint from clothes rod in closet.

Study
1. Adjust door latch/strike.
2. Paint baseboard at north cabinet return.
3. Paint new window sash.

Playroom
1. Door to storage room rubbing on carpet.
2. Repair gypsum board crack at east wall at ceiling.

Storage Room
1. Repair door louver on interior at mitered joints.
2. Install access panel for plumbing valve.
3. Paint attic scuttle trim.

End of Punch List

The punch list, therefore, becomes a comprehensive vehicle to close out construction. Once all items are complete, the builder or contractor is considered to have fulfilled his obligations under the contract. Since this establishes the limits of items to be finished or repairs to be made, it becomes a very important document. Do not let anyone rush you through the final inspection or tell you not to worry about the details. This is your one shot at obtaining the maximum number of corrections. What leverage do you have at this stage? Plenty! The final payout to the builder or contractor should be tied to the completion of the punch list. The building professional knows that in order to receive the last check, the punch list must be substantially completed. It is to your advantage to make the punch list as thorough as possible.

We have reached a time for making a decision. You must decide at what point in construction you will occupy the project. Can you afford the time for the builder or contractor to finish the punch list, or does your schedule demand you move in before all items are finished? This is an important consideration. As soon as you take occupancy, regardless of the level of completion, final payment is usually requested. As construction professionals see it, if you are able to use the final product, they are entitled to their money. Making your last installment payment removes much of your leverage in dealing with some of the stickier items on the punch list. For example, getting the contractor to replace a scratched doorknob will probably be more difficult if he has the final payment in his pocket. The longer you can delay occupying the project, the better.

Many builders or contractors know you are itching to move in. Either you have no other place to live in the case of a new house, or you are sick and tired of the inconvenience of an addition or remodeling. Anticipating this scenario, you may notice construction slowing down near the end. If this is the case, the construction professionals may be trying to get you to occupy the project so they can close it out. Often,

homeowners are at their mercy, having little choice but to accept the new or remodeled house and move in. To avoid this problem, you can insist that as much construction and as many adjustments as possible are completed well before you must move in. If your construction professional tries to postpone fixing outstanding items until the punch list stage, counter with a request to address the problems as soon as they appear. Make it clear that you won't allow necessary construction and needed adjustments to be left until the last minute.

A good strategy here is not to wait for the final walk-through inspection to start compiling your punch list. Attempting to spot every problem during one tour is tough, even for the pros. Begin your own informal list well before construction nears completion, noting the unfinished work or defects that require repair. You will be surprised how certain flaws in finishes can only be seen in certain lighting conditions. Avoid doing punch lists after dark. You want lots of light!

Using your set of partial lists compiled during construction will aid you in being thorough during the final inspection. Share your lists early on with the builder or contractor so he can get a leg-up on addressing these problems. As each item on the punch list is accomplished, cross off the entry. Often new details requiring attention will crop up and should be added to the list. Your professionals should be preparing their own lists and consolidating older lists into shorter documents. Once the punch list line items are eliminated, your project is complete!

Occasionally, resolving certain punch list issues requires you and your project team to do some compromising. Entries that you or your architect consider a defect requiring attention could be regarded as construction within building industry tolerances by the builder or general contractor. As an example, I have friends who had recently completed a kitchen remodeling. After dinner, our hostess showed me what she considered flaws in her new granite countertops. Under a magnifying glass, she pointed out a small discolored area. She asked my opinion. I told her that under normal conditions, without the magnifying glass and her flashlight, I really wouldn't notice the problem. She scoffed at my reply. "Oh, you guys all stick together. My builder said the same thing." She went on about the stone supplier telling her that granite was a natural material and that variations are an integral part of the material. "He actually said, 'It's the way it comes out of the mountain!' " Since both parties were unwilling to bend, it was an old-fashioned standoff.

When a situation reaches the standoff stage, either one side has to give in or a compromise must be reached. The following are some techniques to reach a solution. First, invite an independent, third party with no interest or affiliation with any party in the project to render an opinion. This could be a neutral builder, contractor, subcontractor, or architect. After agreeing on the chosen umpire, all parties should honor the final decision. A second strategy would find a way to compromise. If several punch list issues are being disputed, find a middle ground for all involved. Accept one item as is, in return for having another item fixed. For example, you agree to live with some ceramic tile joints that are not completely square, and the builder will fix the minor blemishes in the drywall. Do a little horse trading! A final technique would find you accepting the installation in question, but receiving a negotiated reduction in the project's final cost. My friend with her "flawed" granite countertop resolved the deadlock by accepting a few hundred dollars off the price. I can just imagine her builder gladly handing over this "hush money" in exchange for some peace and quiet!

Analyzing Warranties

During our discussion about evaluating potential builders or general contractors, we mentioned the importance of checking out their warranty program as part of the selection process. Let's review several issues that may confront you when you are handed a warranty document you can't understand. These programs come in many shapes and sizes. Some are fair to the homeowner. Others can put you at a decided disadvantage.

The best warranty you can receive is a short, simple statement that any faulty materials or workmanship will be corrected for a period of one year from the time of occupancy. This is the same language contained in the AIA contracts we reviewed in chapter 5. If your contract has similar language, you can rest easy. That's all you need. On the other hand, for those homeowners who are handed a multi-page document, the following analysis could come in handy.

In my experience, warranty programs on custom projects are more favorable to the homeowner than those offered on houses purchased in predesigned subdivision developments. With a custom house, you are working with a builder or contractor on a start-from-scratch new home or addition. There's a good chance in this case you'll get a good warranty program. Homebuilders who produce hundreds of houses within a given development, however, are commonly the experts at restricting your remedies. The following examples are clauses you should be familiar with in case you see them in a prospective builder or contractor's warranty.

Limited warranties usually are divided into three sections—terms and conditions, specific exclusions, and a list of physical standards. Terms and conditions first address the specific circumstances under which the construction company will be responsible for problems. Watch out for clauses which mention any of these topics. I have paraphrased the clauses to shorten them.

• *The homeowners must identify defects during a limited number of walk-throughs during the warranty period. Only items discussed during the pre-occupancy inspection shall be included.* This clause puts too many limits on your opportunity to discover any problems. Insist on a full year to discover all problems.

• *The homeowner releases the builder from any damages caused by any breach of the warranty by the builder.* This means that problems that are a consequence of the builder's lack of action on warranty issues are not his responsibility.

• *This warranty is the sole warranty for the project in lieu of any other judicially created rights.* By accepting this statement, you have signed away any rights afforded you under state law. Consumer protection laws have been created by some state governments to protect your interests if a problem occurs. It is important for you to retain these remedies.

• *In the event of a warranty claim, the builder has the sole decision whether to repair, replace, or pay the homeowner reasonable cost of repair or replacement.* This clause enables the construction professionals to establish the cost or value of a claim. By artificially setting the reimbursable cost to you below the actual cost to make the repair, the builder/ contractor could require you to foot the bill for a part of the remedy.

• *The homeowner waives the right to any legal action against the builder/contractor regarding any defect in the construction.* Permitting the inclusion of this clause cancels your rights to use the court system to seek compensation through legal action against the builder for defective work.

The following clauses can usually be found under an exclusion section, which lists items not covered by the warranty program:

• *Any defect which does not result in actual physical damage or loss.* This statement excludes a lot of territory because a problem that does not result in actual deterioration or require replacement is not covered. By this reasoning, a switch that does not activate a light fixture would not be covered because there is no physical damage or replacement loss. The

only thing wrong is that the wire is probably loose at the connection point!

• *Personal property damage or bodily injury.* If any member of your family is hurt through defective construction or your personal property within the house is damaged, this event is not included within the warranty. For example, suppose the garage ceiling falls on your car, causing damage. The warranty will repair the ceiling but not fix your car.

• *The cost of shelter, transportation, food, moving, storage, or other expenses associated with relocation for living accommodations during repairs.* In the rare event that a repair should require you to vacate the house, all costs for maintaining you outside the house are yours.

• *Consequential damages caused by a defect but is not a defect itself.* Assume the garage door comes off the track and causes the wood door molding to be nicked and scratched. The warranty will fix the faulty garage door but not the damage it caused to the door trim.

The final section of a builder's limited warranty deals with an established set of physical standards for identifying specific problems to be covered by the warranty. These statements usually relate to behavior of construction materials in relation to conditions that will require repairs. Here are just a few examples, followed by the builder's explanation.

• *Cracks in concrete.* Cracks are a common feature in concrete foundation walls and floors and should be expected.

• *Walls that bulge or bow out of plumb.* All frame walls have differences and will be repaired if the bow or bulge is more than [a number of inches specified by builder].

• *Floor squeaks.* A squeak-proof floor is not guaranteed and squeaks are not evidence of a defect.

• *Roof leaks.* The roof will not leak except in the event of build-up of snow and ice. The homeowner is required to remove excessive amounts of snow and ice to prevent this occurrence.

• *Exterior trim workmanship.* Separation of joints will be repaired only if the opening exceeds 3/8 inch.

• *Wall leaks due to caulking shrinkage.* All caulking shrinks, and its replacement is not covered under the warranty.

The list continues, setting standards for the amount of cracking, warping, or movement required to be considered a defect. All of the above examples stretch the limits of acceptable workmanship, enabling the builder to avoid making repairs which would ordinarily be required. Many of these issues may be difficult for you to judge if they are too restrictive. Unless you have professional advice to guide you through this maze, you must make the tough decisions. Your best recourse reverts back to checking the builder or contractor's references. Did they repair all defects or use their custom warranty as an excuse? Were the repairs made on a timely basis, or did past clients have to continually call or make threats? These are important questions to ask, especially if the warranty of the prospective builder or general contractor seems too restrictive.

To ease your concerns a bit, it has been my experience that although reputable builders may have restrictive written warranty programs, they will often exceed their responsibilities by repairing many items supposedly not covered. I constantly hear stories about contractors or builders going the extra distance to please their customer and protect their own reputation. As we have discussed several times, hiring the right professionals is the best method for providing quality at all stages of your project.

How to Make the Best Use of Your Warranty

Since warranty programs usually only run a year, you should have a strategy for obtaining the best results and attention during this limited time. After the final punch list has been completed, start compiling a new list of items, which require attention. The story at the beginning of the chapter is similar to that of the little boy who cried wolf once too often. Daily calls to the builder or contractor for single problems will encourage them to ignore you. Unless you have an emergency, package a list of several items together and call every few months. Construction professionals prefer to have tradesmen return to repair several problems at the same time. Their time is valuable and they will be more accommodating if a trip to your house can accomplish the majority of issues in one visit.

When requesting warranty service, act in a polite and professional manner. There is no reason to lose your cool if the professional honors the warranty. It usually is a good idea to put adjustments you request in writing. This establishes a clear date that you requested service and notes each item requiring attention. This documentation could come in handy if you experience problems obtaining prompt service. Written proof that you informed the construction company in a timely manner could head off future disputes concerning when an issue first occurred. As a rule of thumb, call immediately if a condition can cause damage to surrounding construction, such as a dripping pipe or roof leak. Bundle together problems like drywall defects, sticking doors or squeaking floors that cause inconvenience, but no damage.

The most effective use of your warranty comes at the twelfth-month interval. Your warranty period is almost expired and this is your final opportunity to get repairs made. Consolidate your previous lists into one comprehensive document, and call your construction professional for a twelfth-month checkup. Review all outstanding items and obtain dates from the representative for final repairs. Just as you carefully walked through the completed project to assemble the punch list, spend the time carefully exploring the year-old construction. This is your final shot, make it good!

If during the warranty period, your builder or contractor does not give you the attention you deserve, there are very few strategies you may pursue. After phone calls and letters go unanswered, or no one shows up for scheduled appointments, you can begin to feel powerless. Your leverage to make the construction professional adhere to the warranty is long gone because the final payment has been made. Unfortunately, your only option to enforce the warranty is legal. Having an attorney draft a threatening letter referring to the requirements of the contract can often do the trick. Hopefully, your project will not end up in this category. Again, contacting references to check the performance history of prospective builders or contractors during the warranty period is of key importance. Most construction professionals seem to view warranty work in the same way. Either they consistently make a practice of honoring their commitments by giving good service, or they routinely shirk their responsibilities. They always seem to fall into these two distinct categories. Once they find a niche, they never change! The most effective warranty for you is to hire those with the best service track record.

After the Warranty Ends

What happens after the warranty ends? If construction problems occur, often they appear after the year warranty period is up. Let's first discuss what constitutes a post-warranty problem. Don't confuse maintenance of a house with construction problems. All houses require normal repairs over time. Furnaces must periodically be serviced, pipes can become clogged and doors can swell during hot weather. Problems such as these relate to maintenance and not poor construction. Anticipate that every house requires regular upkeep to preserve its good condition, much like a car. Let your property run down and worse problems are bound to occur.

You should also be aware that certain items installed in your project carry separate warranties longer than the one-year construction program. Appliances, cabi-

netry, interior finishes, windows, and roofing are just a few guaranteed to perform by their manufacturer for a period of years. If you experience problems with your roofing shingles five years after construction concluded, your builder or contractor will probably refer you to the manufacturer for service. The rule in these cases is materials and equipment may carry longer warranties, but workmanship does not. If a problem relates to faulty materials, the manufacturer will usually stand behind their products. Poor workmanship or materials not installed in accordance with instructions will find the manufacturer referring you back to the installing building professionals.

Before the warranty period begins, I highly recommend that you get a comprehensive list of material suppliers and subcontractors. You should be given a tidy package of owners' instruction manuals and warranty cards to fill out and mail in. These are the direct manufacturers' guarantees you are entitled to receive. On a new house or substantial addition project, you would be surprised how many of these packets will be coming your way. Take the time to complete and send the forms; they won't do any good sitting in the instruction book when a problem occurs later!

Figure 12-2 illustrates a list of suppliers and subcontractors. Entries include key subcontractors such as the electrician, plumber, and heating and air conditioning company. Suppliers of building materials are also listed for doors and hardware, roofing, windows, and cabinetry. If a problem in construction occurs after warranty, this list gives you the information required to find a way to address the issue. Identifying suppliers and installers can pay big dividends if problems persist, as you can go right to the source. Your first call should be to the builder or general contractor. Let him assess the problem and contact his subcontractors. Down the road if your construction professional disappears, call the specific subcontractor for attention.

Figure 12-2

List of Subcontractors and Suppliers

Trade	Company Name	Address	Telephone #	Contact Name
Excavation and Grading	Deep Excavating	Somewhere, USA	(555) 555-5555	
Concrete	Durable Construction	Somewhere, USA	(555) 555-5555	
Site Utilities	Acme Sites	Somewhere, USA	(555) 555-5555	
Steel Fabricator	Strong Stuff, Inc.	Somewhere, USA	(555) 555-5555	
Carpenter	JM Construction	Somewhere, USA	(555) 555-5555	John Moore
Door and Hardware Manufacturer	Knobs Are Us	Somewhere, USA	(555) 555-5555	Frank Schneider
Window Manufacturer	Glass, Inc.	Somewhere, USA	(555) 555-5555	
Roofer	Tops Roofing	Somewhere, USA	(555) 555-5555	Neil O'Connor
Wood Component	Truss, Inc.	Somewhere, USA	(555) 555-5555	
Insulator	Smith Insulation	Somewhere, USA	(555) 555-5555	Thomas Smith
Plumber	Pipes To You	Somewhere, USA	(555) 555-5555	
Electrician	Lake Electric	Somewhere, USA	(555) 555-5555	Robert Lake
Heating/Air Conditioning	Reds Heating	Somewhere, USA	(555) 555-5555	
Gypsum Board	JM Construction	Somewhere, USA	(555) 555-5555	John Moore
Painter	Splash It On	Somewhere, USA	(555) 555-5555	
Cabinetry	R & Q Cabinets	Somewhere, USA	(555) 555-5555	Rob Quinn
Flooring	Floors, etc.	Somewhere, USA	(555) 555-5555	
Ceramic Tile	Floors, etc.	Somewhere, USA	(555) 555-5555	
Gutters	Tops Roofing	Somewhere, USA	(555) 555-5555	Neil O'Connor
Garage Door	Ups & Downs	Somewhere, USA	(555) 555-5555	
Drive Paver	Smooth Drive Co.	Somewhere, USA	(555) 555-5555	
Appliances	XYZ, Inc.	Somewhere, USA	(555) 555-5555	Dick Young
Fireplace	Warm Glow	Somewhere, USA	(555) 555-5555	
Landscaper	Green Meadows	Somewhere, USA	(555) 555-5555	

An additional solution is to consider extended construction warranty programs. These are policies issued by insurance companies. They work similarly to extended car guaranty programs. They are available from builders, Realtors, or even on the Internet. Programs run in different time intervals and usually cover internal systems and the structural integrity of the house. Internal systems such as the electrical, plumbing, heating and air conditioning systems, and appliances can be covered for a number of years. Specifically, these policies cover faulty electric wiring, plumbing piping and the heating system, guaranteeing they will be repaired. The structural system, consisting of the foundation, columns, beams, bearing walls, and roof framing can be insured up to a period of ten years. The costs vary, and like all insurance policies, can carry deductibles and service call fees.

Just like construction warranties furnished by your professional, these policies also carry restrictive terms and conditions along with exclusions. If you are interested in this additional protection, read the fine print as you would on any other document. Are there programs worth the money? You probably already have experience with similar policies when you have decided upon service programs offered with appliances and cars. While some are worthwhile, the expense of others may outweigh the advantages. This is a calculated risk. If you have confidence in the construction companies you hired, along with their materials and workmanship, this additional coverage may not be necessary. On the other hand, if you have an uneasy feeling about how well your project was constructed, you can have the peace of mind from this additional insurance once the project is completed. Verify that the extended warranty company is financially sound and will be around to honor its commitments!

Congratulations! By now you should be enjoying the benefits of your project. All your hard work should have paid off if you followed our six-step process. I have tried to point out the key areas that require your attention and effort. I hope I have not scared you into saying, "It's got to be easier to just buy a house than going through all that hard work!" If you read the first few chapters, understand the process, and familiarize yourself with how the key players work, you should be in a good position to hire building professionals who will serve you well. Any potential problems should not become stumbling blocks if you follow the chapters step by step.

The last part of the book addresses issues specifically related to additions and remodeling projects. Although many of the principles are the same as we learned for a start-from-scratch project, there is additional information worth knowing if you plan to expand or remodel your existing house.

Figure 12-3. The finished project.

Chapter Twelve Recap

- Your goal is to insure that all construction required by the contract is either complete by the time you occupy the project, or scheduled to be accomplished soon after.

- Working with your professionals, develop a *punch list*, a document that comprehensively lists every item to be completed on a room by room basis.

- Don't wait for the final inspection to start your list. Trying to see everything on one walk-through is difficult. Share your lists with your professionals to establish items requiring correction or completion.

- If agreement cannot be reached on certain items you feel require correction, consider having an independent third party render a binding opinion. Reaching a compromise by finding a middle ground through bargaining on items can also be effective.

- Once you occupy the project, the construction warranty begins. It is very important that your warranty furnished by the construction professional is not unfairly limited through exclusion of specific problems.

- Bundle your warranty items in groups and only call for corrections when several items have been accumulated. Builders and contractors will give you much better attention using this method rather than calling daily with single issues.

- Consider extended warranty programs from insurance companies to supplement a contractor's warranty program that may be too limited.

Part Three

Remodeling and Adding On to Your House—What You Need to Know and Do

Chapter 13

Remodeling or Adding On—the Best Bet for Your House

Remodeling and home addition projects present many issues that are unique in comparison to start-from-scratch construction. To produce a successful project of this type, you'll need to take many of the same steps in hiring a building team and soliciting bids described in earlier chapters. But certain limitations you will encounter with remodelings and additions require different activities and solutions.

New House/More House will show you how to gather important information, address key planning problems, and obtain inexpensive professional advice and expertise. You can evaluate many of these issues by conducting your own preconstruction study or you may prefer to hire building professionals to provide this analysis. Either way, you'll gain information at relatively little expense, which will help you decide whether to go ahead with a remodeling or addition project.

If you need more living area or want to enhance the space you now have, remodeling or adding on to your present house may be the answer instead of picking up and moving. In many ways, these projects require a different way of thinking and are often much more challenging than new house projects. You may find yourself working in tight quarters within predetermined walls, or dealing with land constraints. Be-

fore you move ahead with a remodeling or addition, I recommend you first look into some concerns which are unique to this project category and may have not occurred to you. Stick with me while we explore the possibilities!

Financial Issues

Any remodeling or addition project is first a financial decision that carries several consequences. Do the real estate home values in the neighborhood or area warrant this additional investment? If you had to sell your revamped house, could you recover the total cost of the original house plus the improvements? There is an old maxim in real estate, "Don't be the most expensive house on the block!" You must determine if your plans for a project would price you out of the local real estate market. The following discussion offers a few examples for your consideration.

Whether you live in a defined subdivision or a neighborhood that has loose boundaries, the mix of existing house and lot sizes establishes a maximum price someone would be willing to pay to live there. For example, if the average cost of a house in a given area is $200,000, few people will be interested in paying $350,000 for a house, no matter how wonderful an addition turned out. For that kind of money, they will seek out another area where the average home price is closer to this higher figure. An exception to this rule occurs in older neighborhoods in metropolitan areas where renovations and "urban homesteading" is popular. If they wait long enough for others to invest in a trendy area, homeowners can cover their financial risk and often realize significant gains.

Taking our example in the opposite direction, a house valued in the lower tier of neighborhood prices may be a good investment. Popular areas will attract homeowners who will buy a smaller house in exchange for living in a more prestigious property. The increase in values from the larger neighboring properties will pull their value upwards. They may in turn decide at some point to improve the house through an addition or remodeling because the area warrants the additional investment. Taking this one step further, a recent phenomenon finds homeowners buying smaller houses in upscale areas for the purpose of tearing down the old structure to build new. They are essentially buying an expensive vacant lot.

To find out whether a certain area will support the increased cost of an alteration or renovation, an examination of the local real estate multiple listings or a chat with a local Realtor can quickly provide this information. Realtors usually do not mind spending time discussing these issues with you. They see it as an opportunity to sell you a new house, as well as selling your present residence if you decide an expansion project is not in the cards. This expert information is invaluable and is usually given with no obligation.

If you decide to be the first on your block of ranch houses to add a second story, hope and pray that you will set a trend for others to follow! Becoming the pioneer in a neighborhood requires a calculated risk and a commitment to stay for an extended period of time, until others follow your lead. This brings up another concept that may enter the picture—*utility*. I have participated in several projects in which

homeowners decided to buck the trend and remodel or build on to their house anyway, despite an unfavorable financial analysis. The usual reason is that they love the area and wish to stay for their neighbors, their children's circle of friends, and schools.

They rationalize that although the current real estate values may not be favorable, they plan to stay for most of their lives and will gain the pleasure and utility from the addition or remodeling project. If they stay long enough, they gamble that others will similarly add money to the area and their investment will be justified. This decision-making process should be weighed against the anticipated future of the area. Neighborhoods that are on the ascent are a safer bet than communities that are witnessing no increase in real estate values.

Get as much advice as you can find before committing your funds. Lenders and financial institutions can be the deciding factor. When you seek a building loan for a project of good size, an appraisal of your present home and the proposed improvements will be required. The total cost of the two combined will be professionally compared to prevailing present and future real estate values. If the cost of your finished project exceeds projections of future real estate values, your loan request may be declined and the decision is out of your hands.

Visit City Hall

When considering an addition, your first job is to determine the amount of land available on your lot for expansion. Using laws known as zoning codes, every county or municipality allots a certain specific site area for a house to occupy. These codes also regulate the size and height of construction. When I first visit a homeowner interested in an addition, I know my very first task is to examine a survey of the lot and find out whether any additional space is permitted. Many a potential addition has been quickly scaled back or even eliminated from consideration as a result of this quick analysis.

You can perform this evaluation quite easily on your own. When you purchased your house or even refinanced the mortgage, a survey was prepared, showing the physical boundaries of your lot and the precise placement of the house. Once you have this drawing, a call or trip to the local building department is the next step. Can't find the survey anywhere in your document file? No problem. Your building department may be able to help you. When you arrive at the local office, tell them the purpose of your visit and ask to obtain all the pertinent zoning information for your lot.

Using your survey or a map book from the building department, look for the *setback lines.* A setback is the distance measured in feet from the boundary lines of your lot to the exterior walls of the house. The zoning code establishes a front, side and rear setback distance for every lot. Many lot surveys will also show the front setback, but we are primarily interested in the side and rear measurements where most additions occur. If your lot is a rectangle consisting of four lines, use your survey or map book to place a line for each setback. This will produce a smaller rectangle as shown in figure 13-1 (page 178). This smaller area is the allowable limit for construction. Looking at our example, you will notice that the corner lot shown at the intersection of the two streets has only a little room available for expansion on the right side and the rear. The next lot above the corner is known as an interior lot and has ample space available behind the existing house.

If you find that little space is available for any expansion, your project plans may have to be adapted. In this case, there are two alternatives to consider, adding another story or applying for a zoning variation. Before we cover these options, we must discuss a few more potential zoning issues.

Most zoning codes also control the amount of floor area permitted in proportion to the size of the lot. You may find that although the setbacks present no problem, the code further requires that the first floor be no more than a certain percentage of the lot area. For example, if your lot size is 10,000 square feet, and the allowable coverage is 20 percent, that would limit the first floor to 2,000 square feet. *Floor Area Ratios* (FAR) are part of every zoning code and restrict the total area of the house.

Even more extensive zoning restrictions are established in certain neighborhoods. *Bulk* ordinances control the amount of three-dimensional cubic area allowed within a house. The concept behind these rules is to discourage large new houses or additions being built in residential areas of older small houses and lots, thus preserving the original scale of the properties on each street. As I mentioned, in upscale areas, it is now very popular to tear down a small house and build a bigger new house that extends to the limits of each setback line. The houses typically have two-story or high-vaulted spaces, which increases their height. Bulk ordinances are used to combat this trend. If this is your intention for your project, you may need professional help to understand the required complex zoning calculations.

Another zoning requirement gaining in popularity is the preservation of trees. Communities that have large stocks of mature trees have enacted ordinances requiring permission to cut down existing trees for building projects. Many towns now view trees as a natural resource worth protecting. Cutting down mature trees can change the appearance of an area in a single day. Unfortunately, new trees planted to replace removed trees won't grow overnight.

When I visit older neighborhoods for a potential addition project, my heart sinks when I see a big oak tree right in the area where the addition would be built. Not only is the possibility of the project in question, but I feel guilty at the prospect of destroying a tree that took so long to grow. Municipalities with these codes will occasionally permit the removal of existing trees in exchange for the planting of new trees matching the equivalent size. For example, if a

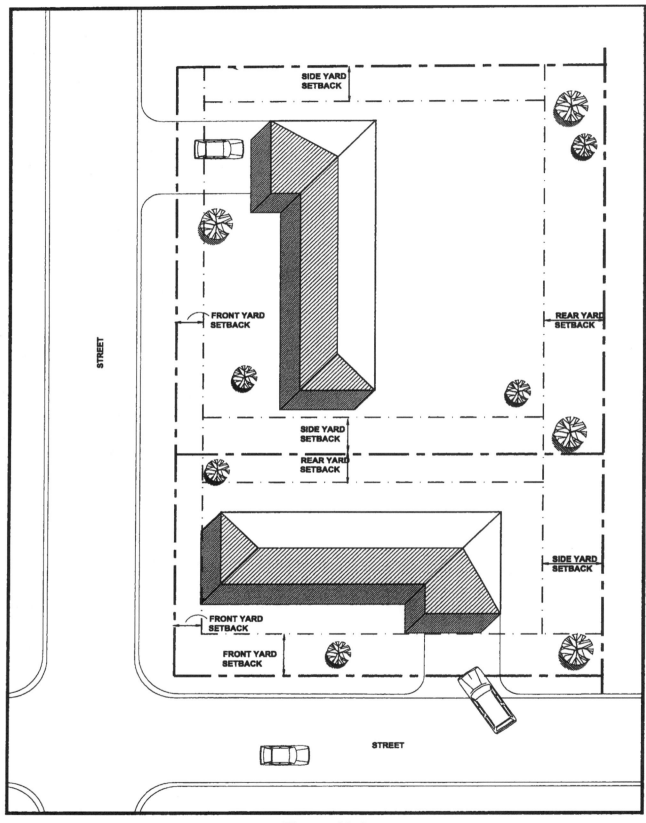

Figure 13-1
Site Plan Illustration

fifteen-inch diameter oak tree is to be removed for an addition, three five-inch diameter trees will be planted around the addition to compensate for the loss. Each town has its own requirements, which you should determine before making construction plans.

Zoning Variations

If your potential addition project encounters one or more of these restrictions, a process known as a zoning variation may be available to you for relief from these limitations. All counties or municipalities have zoning boards of appeal set up for this specific purpose. They are composed of volunteer citizens from the local area and hold hearings every month. Acting as an advisory group, they conduct a public meeting to weigh the merits of each case and forward their recommendation to the city council for final action. As a rule, zoning boards of appeal can consider granting variations from 20 percent to 30 percent of the original requirements. This would allow you to build from 20 percent to 30 percent closer to the lot line than the ordinance states. In the case of a rear setback requirement of thirty feet, the zoning board is empowered to recommend a reduction of six to nine feet, depending on their specific guidelines. Thus, a board with a 20 percent protocol could grant a variance to reduce the rear setback from thirty to twenty-four feet.

A zoning appeal follows virtually the same routine for any county or city. You must make an application explaining your hardship and the amount of variance you seek. A moderate fee accompanying the application is usually required. Depending on the complexities of the project, a minimal architectural site plan may be required to help the board better visualize your proposal.

Zoning board meetings are a great example of democracy in action. You or the local authority is required to notify adjacent property owners of the variation request through registered mail. A brief description of the request is also sent along with an invitation to attend the hearing or visit city hall to review the application. A public notice may also be advertised in the official announcements or classified section of the local newspaper. To ensure even more public notice, a sign proclaiming the public hearing date is often placed prominently on the property.

I have attended countless zoning variations as a representative for my clients, making as convincing a presentation as possible to highlight the merits of each case. Since the dynamics of every board and the surrounding neighbors vary, the course and outcome of each hearing is different. I offer my clients the following advice gained over the years in preparing for these hearings. Before the hearing is scheduled, homeowners should informally contact their neighbors to acquaint them with the specifics of their request for a variance. Obviously, your immediate neighbors to the sides and rear will be most affected, but neighbors up to 500 feet away will probably also be invited for their input. Although they may not be immediately impacted, the variance could set a precedent for the entire neighborhood. Generally, only neighbors who take issue with your request will attend the public hearing. Homeowners who have no problems with the proposal seldom appear.

If you discover that any of your neighbors do object, address their problems before the hearing. For example, let's assume you are seeking to build a family room addition onto the back of the house and need a five-foot variance. However, your neighbors to the back don't want to look at any more house from their rear window. You should try to find a compromise to gain their support. In exchange for the extra area, you might offer to plant additional landscaping to soften the impact of the addition. Your goal is to turn negative voices that will protest at the hearing into positive supporters who will stay home. Zoning hearings permit anyone attending to ask questions or state opinions. Board members will frequently take into account negative comments that have merit when making their decision.

The hearing begins with a presentation by the homeowner or his building professional, if he are represented. Your most effective approach is to demonstrate a compelling reason for needing more space. For example, you are expecting another child, and not wanting to move because you love the area, you need the variance to accommodate the new family member. In my experience, there is nothing more effective than a noticeably pregnant mother present at the hearing. It is very difficult for board members to tell her no! Highlight your solutions to make the project acceptable to your neighbors by going the extra yard to beautify the exterior. If past variations in the area have been approved, use these precedents to strengthen your case.

While considering all the opinions given at the hearing, board members may propose their own compromises to make everyone happy. Be prepared before the meeting to know your fall-back position—the minimum variance you are willing to accept. Remember, the zoning board's actions are usually only advisory. If you get a positive recommendation, the city or county board must also review the request during a regularly scheduled meeting and make the ultimate decision. Occasionally, cases denied by the zoning board can be overturned and granted by the higher authority. The opposite can also occur, although a positive zoning board action is rarely reversed by the city council.

Outcomes of these hearings vary. I am often surprised when variances that I thought had little chance are granted, while cases I thought were slam-dunks are denied. As with any public governmental proceeding, you roll the dice, take your best shot, and hope for the best!

Analyze Your Lot's Potential

Every house property has existing features that will present either advantages or drawbacks for an addition or remodeling project. These include limitations due to underground utilities, the condition of the house, and aesthetics of the exterior and interior. Each condition must be analyzed, since it may determine whether enlarging or remodeling the house is feasible. You can begin to conduct this study on your own, identifying conditions like those in the following discussion that could be potential problems. Final analysis of your lot's potential may require consultation with a building professional, who may also be able to help solve some of the problems. But keep in mind that certain challenges may stop a project in its infancy, as we pointed out in the discussion on setbacks. Other problems won't reveal themselves until construction is under way. By summing up the range of these challenges, you can decide whether the project you envision has merit.

The first step is determining how friendly your lot will be for an addition. Underground utilities such as water, sewer, electric, telephone, natural gas and cable television services are commonly buried underground. Most communities offer a free location service run by a joint commission of utility companies that is just a phone call away. These services have various names such as J.U.L.I.E. or D.I.G.G.E.R. Each utility will come out to your property and identify underground lines by either spray painting the ground or leaving a line of little flags. Within forty-eight hours you can learn whether existing utility connections lie within your potential addition area. If you have a well or septic field, your best bet is to contact the local county health department to check its records for past plans indicating their exact positioning.

Relocating underground utility lines can be very expensive. In general, utility services involving piping, such as water, sewer and gas, are far more costly to reroute than those installed with wiring, like electrical, telephone and cable. Relocating a septic field is the most difficult. Unless you have additional land, there may not be enough area available to meet local health department regulations. Making the decision to relocate these utilities, if physically possible, is budgetary in nature. If your project budget can't incorporate the cost, you may want to consider moving to gain more space.

The contour, or slope, of your property can also play a part in determining the feasibility of an addition. Land that drops sharply away from the house makes it difficult to maintain the same floor level in an addition unless you introduce either steps down or a lower level. In areas of the country where basements or crawl spaces are common, this is the usual solution for steep slopes. Geographic locations without this option can use stilts. Building on sloped land is usually more expensive, because of the need to accommodate the height difference.

Finally, your analysis to decide whether your project is feasible should also include existing landscaping. Large trees and mature plantings will either have to be eliminated, or if possible, transplanted. This is another budget and time consideration. Establishing new trees and gardens require a lot of effort and expense, not to mention the time required for trees to grow in height. Transplanting existing landscaping is also risky, as not all plants tolerate a new location. If existing trees and gardens are a hindrance to your addition, add money to your budget to address this issue.

Analyze the Potential of Your House

Let's turn to the subject of your existing house. Adding to or remodeling an older home may be made more difficult by a set of problems inherent in the home's condition. Depending on the age of the house and how well it has been maintained, some of these conditions may be hidden and can cause unfriendly surprises when exposed during construction. The following is a list of these culprits and the resulting consequences.

• Electrical wiring and equipment is defective. Older wiring inside walls should be suspect, because the original insulation can rot away. This is an unsafe condition that requires removal and rewiring.

Electrical panels, especially the fuse type, are antiquated and no longer permitted by electrical codes for new construction. Updating an electrical service panel can cost thousands of dollars.

• Cast iron or clay plumbing pipe in older homes can deteriorate over the course of many years. If you expect to reuse existing piping, there is no worse surprise than opening a plumbing wall only to find badly corroded pipes requiring replacement. This can be a nasty financial shock when you least expect it.

• Rotted or insect-laden wood can often be exposed in areas that have little ventilation, such as crawl spaces, attics and walls or floors adjacent to plumbing piping. Since you cannot build good onto bad, replacement is the only option. This scenario can also create bombshells for your budget.

• The incoming utility services can be too small to handle the heavier load. Additional bathrooms typically require a larger incoming water service, as most houses are not constructed with surplus capacity. Electrical services are also a target for early examination as the number of available circuits may be limited or nonexistent. If your property utilizes a septic field, adding bedrooms or bathrooms will require extending the field, provided you have the additional land available. Dealing with these problems adds thousands of dollars to the project cost.

Occasionally, opening up walls will reveal original construction that doesn't meet current building codes or never met any code at all. The old saying "They don't build them like they used to" doesn't extend to all houses. After working on countless remodelings and additions to older properties, my usual reply is, "Thank God they don't build them like they used to!" If existing construction, particularly plumbing and electrical work, has code violations, the local building department will likely require alterations to bring them into compliance.

These are just a few of the leading cast of characters you can encounter on projects of this type. My main purpose in discussing these problems is to alert you to budgeting concerns. If you are planning to remodel an older house or one that has been poorly maintained, plan on adding at least 10 percent of the budget to cover these unexpected surprises. This will make the funds immediately available for repairs, instead of forcing you to scrounge for more money.

Consider Appearance and Function

The next problems concern aesthetics, or the appearance of your finished project. Altering an existing house's interior or exterior style requires careful consideration. Let's begin by discussing the exterior challenges. If you are thinking about building an addition, how will the new exterior affect the outward appearance of the house?

Reduced to the basics and depending on the size of the project, you have two choices. The original exterior style of the house could be continued, making the enlarged structure a natural extension of the original design. This is the most common approach and works well if you can find new exterior materials that match up with the original. Unfortunately, bricks from a particular supplier are never made for more than a few years, since the clay, which is dug from the earth, is soon exhausted. The same is true of stone, as quarries close, the supply for a particular stone ends. Windows and roofing shingles also see frequent manufacturing changes because styles and colors come and go. Often, additions will require a complete new roof for just this reason.

The second choice for the homeowner is to adopt a new style on the addition and rebuild the exterior of the original house to match. If you are dissatisfied with the way the house now looks, adding more of the same may not make good sense. A large addition can provide the opportunity to give your house a makeover or face-lift. I feel an addition is successful if a casual observer cannot tell where the old house ended and the extension began. There is nothing worse than an obvious addition in a completely alien style. It is as if another part of a house fell out of the sky and happened to attach itself to a particular home! I strive to make my expansion projects look as if the entire house was built at the same time.

Just as the exterior of an addition must match the original, so must the interior of the expanded house. Remodeling and addition projects frequently alter the design style for the interior, since tastes in color and form now change so quickly. Designs that are just a few years old when a new modification is planned can frequently be out of style. How much of the existing finishes and colors will you have to change to blend the new with the old? This is a hard question to answer. Some new styles can adapt to older looks, depending on your personal taste. If the new project is contained in more isolated areas such as master bedroom suites, bathrooms or kitchens, tolerating different design styles may be acceptable. Changes in color or style are not often objectionable if they are not immediately adjacent to each other.

If the project will change at least half of the house, be prepared to alter the entire interior. Unlike the exterior, which has permanent materials, interiors can be modified over the course of time. Instead of changing everything immediately, you can replace or rework the original elements one phase at a time. Depending on what you can afford, you can do a little updating each year until you are satisfied with the final look.

Another important consideration is how building an addition or making other changes to your house will affect the way it functions. Remodeling and addition projects often face challenges centering on accessing new rooms and scaling their size to match the rest of the house. When adding rooms onto a house, avoiding strange paths for circulating from one space to another can be very difficult. Designers refer to circulation as *flow*. Awkward transitions from one room to the next break up the house's continuity and call attention to the addition. Long, twisting hallways built to reach new bedrooms are a prime example. Because of the inconvenient layout of existing rooms and the limits of expansion space, I have often advised prospective clients against an addition project. No matter how hard you try to solve the puzzle, certain houses are simply not good candidates to be expanded, because the existing spaces will not present good flow into the new.

The size of new spaces should be proportioned to fit the original house. If you are planning to add larger, open rooms such as family rooms or great rooms, will they dwarf the smaller spaces in comparison? It can be awkward to walk through a house with small cramped rooms to gain access to a huge new open space. The original rooms will be made to appear even smaller when matched with the new kid on the block. I have visited many houses where the addition satisfied the needs of the family, but the home had two identities. Old space was cramped and confined while the new area was large and spacious. Designing an addition or remodeling project is an exercise in balancing the nature and size of the old with the new.

Should you want to take a stab at laying out addition or remodeling plans before contacting a building professional, here are a few pointers. The first step is to have a floor plan of each level available to begin your planning. The original construction drawings are your best bet. If you don't have the plans, check with the local building department. Depending on the age of the house and their record-keeping history, you may find a copy stored away at city hall. Failing this, your only recourse is to take a tape measure and size each room. If you are lucky enough to have plans, double-checking the actual size is still a good idea, as plan dimensions are often modified during construction.

If you are starting from scratch, you will need to transfer your measurements to paper. You can do this by simply using a ruler and some square graph paper to sketch up your floor plans. If you are comfort-

able using a home computer, software programs are available to assist in putting your existing house into a drawing format. Regardless of your drawing medium, try expanding your house with an addition or move existing walls around for a remodeling. While this exercise can provide fun for the whole family, you may quickly discover that architecture isn't as easy as it looks. Some of the walls you need to move may actually hold something up! Although you will ultimately require professional assistance, the process I just described will enable you to start thinking about the possibilities of a project. You may discover you have talents you never knew existed!

Building Upward or Outward

While conducting your research at the local building department, you may have discovered that the amount of land available for an addition was too limited. For those homeowners with single-story ranch houses, adding on a second story is another option that should be evaluated. Building upward has many concerns in common with our previous discussion about design style issues and circulation planning. By extending the house with a second story, you should follow the rules about either preserving the same appearance as the original house or completely altering the style. You would be surprised how adding an extra story can drastically modify the look of the original house. Keep in mind that the best addition should look like it was always part of the original house!

If you have enough land to expand horizontally, making the decision to build upward or outward is dependent on several factors in the present house. First, let's discuss how the addition will connect to the existing house. We have already described the importance of circulation from the old into the new in a horizontal extension. A clear, convenient path must be logically available to access the new rooms. You would prefer not having to reach your new master bedroom suite via the dining room.

This same thinking applies to adding a story. A stairway has to be added where it is convenient for circulation and looks like part of the original house. In a small ranch house, this can be a difficult task. Often there is not a large entry or an extra-wide hallway available to locate the stairs. It may seem the only option is to place the new stairs in the living room area or some other inappropriate location. This is one of the major challenges facing the planning of a successful second story addition. Nothing looks worse than a staircase stuck in the wrong spot!

Another concern in expanding upward is structural. The existing foundation must be able to support the weight of an additional floor level. Plans for building a second story can come to a halt very quickly if the foundation is designed to support only one story. If you have the existing plans, a building professional can check the capacity of the foundation quite easily. If plans are not available, there are other ways to check the foundation. In regions of the country where basements are common, foundations usually constructed with these underground levels have the capacity for a second floor. Houses built on slabs or crawl spaces may not. The only method to check out the shallow foundation system is to dig down along the exterior wall and measure the existing foundation. In every instance, the question of structural support will have to be determined by a professional.

Each approach to expanding your house has different cost implications. Horizontal additions disturb adjacent land, requiring excavation and resloping of the ground. As I previously mentioned, relocating underground utilities in the way of an addition is expensive. New foundation construction is not cheap, either. Often a portion of a ranch house roof must be reworked in order to function correctly with the new roof of a horizontal addition. Adding upward obviously requires the removal of the roof. Existing attic floor framing lumber is usually too undersized to support the new second floor, as building codes permit lighter load calculations for attics. New, deeper floor framing usually must be installed between each original member, or an entire new floor substituted. Additions usually require removing part of your original house before starting to build new construction. Part of the project cost is actually paying to dismantle useful parts of the old house. It is often difficult for homeowners to understand why they must pay to remove something they were previously using!

Addition projects also carry their share of stress for homeowners, as we discussed in depth in chapter 6. This is particularly true for additions, whether upwards or outwards. Horizontal extensions do have one big advantage, as the majority of the work can usually be accomplished without disturbing the interior of the original house. The foundation and exterior wall and roof framing can be built with little interruption. Connecting the utilities from the old to the new usually accounts for some minor inconvenience. Only when the interior of the new space is well advanced will the addition be opened up to the original house. At this point, changes will be made in the original house to "marry" the addition to what's already there.

This advantage does not apply to upward expansions. The first step is to remove the roof, requiring evacuating much of your personal property and quite possibly your family. I have participated in many of these types of projects. There is nothing I have ever seen to match the look of homeowners standing in the middle of their house, seeing only four walls open to the blue sky above!

On one project, the homeowner and I were meeting with the general contractor after the roof had just been removed. As my client looked around at what was once his house, he remarked, "Boy, I sure am trusting you two guys!" Seeing a home without a roof could be compared to a tornado removing part of the house. There is little left! Until the new roof is installed, you are at the mercy of the weather and your builder/contractor's ability to temporarily keep out the rain. I always recommend starting these projects in the dry season!

Since most families don't have the luxury of moving out for an extended period, they must adapt to meet the challenges of daily living in a construction zone. An experienced construction professional should be accustomed to accommodating you while the addition or remodeling project progresses. Temporary walls can be erected to protect parts of the house that are unaffected by construction. Interruptions to utilities can be scheduled in advance to permit you to plan for the inconvenience. I know a lot of homeowners who after living through an addition or remodeling pro,ject say "Never again!" No matter how accurately I try to explain and forewarn them about the disruption to their lives, it usually still comes as a surprise. Patience and flexibility is the name of the game.

Make Your Final Analysis

After researching the issues we have just discussed, you must weigh the information gathered to make a decision. Do you add-on, remodel, or sell the house and move? Frequently, homeowners seek professional assistance to help them make their decision. When I meet with potential clients facing these questions and wishing to assess all the possibilities, I recommend a two-phase planning project.

During the first part of the project, I conduct a short feasibility study to help them make their decision. The second part, which takes the project through the remainder of the process, occurs only if the outcome of the first part produces a favorable verdict. Many homeowners find this professional service package appealing, as it does not commit them at first to an entire project. If the results of the study are unfavorable, they are only responsible for a fraction of what full project design services would cost.

My efforts during the first phase begin with assessing the property. Using the criteria given to you ear-lier in this chapter, I form an evaluation of the positive potential and the negative drawbacks the house presents. Due to some of the issues we discussed, such as limited setbacks, undersized utilities, or structural shortcomings, a project may be halted. But if no major roadblocks are encountered, I proceed into the preliminary design phase. Basic floor plans and exterior elevations are prepared to illustrate how the envisioned project would function and appear. Accompanying the drawings, I prepare a rough construction estimate, usually expressed in a low to high range.

The last component of the first phase study is an anticipated project timetable, describing the length of time to complete the project. Armed with these three documents, I can present my client with all the information needed to make an informed, intelligent decision. We discuss how well the design will fulfill their functional requirements and aesthetic goals. This, in turn, will be compared to the anticipated construction cost, analyzing areas where any possible savings can be made. Finally, the timetable provides a clear picture of whether living arrangements and construction schedules can be maintained without interrupting daily routines. We also discuss ancillary issues, such as whether altering their home justifies the cost and inconvenience of proceeding with the project—or whether they would benefit more by avoiding the inconvenience and spending their money on a different house elsewhere.

At this point, I push my chair back from the table and tell them they have all the information required to make a decision. My clients vary in their responses. Some enthusiastically say, "Keep going, we are committed to begin phase two." Others want to take some time to consider their options. Some homeowners in this category never choose to proceed. Others will take a few weeks or months to pull the trigger. One client actually called me two years later and said, "We thought about it and I guess we are ready to move ahead!" Everyone makes decisions in their own time. Issues such as changing jobs or having more children often come into play.

Whatever the variables may be, I highly recommend that you use this two-phase planning concept. Professional advice often makes the difference in accurately determining whether a project should move forward. A feasibility study commits you to the smallest possible fee and puts you in a good position to make the ultimate decision. Either you are going to make do with the current conditions, sell the house and move to something different, or modify as planned in your design study. In a rare instance, I have torn down the existing house and built a new residence on the same lot while my clients lived elsewhere! If you want to stay in the neighborhood, many ways can be found to fulfill your needs.

Making a positive decision to proceed with an addition or remodeling project means you are going to use the same concepts and principles presented in the first twelve chapters. Most of the issues you will encounter will directly apply to this type of project, no matter how small. Even a basic kitchen or bathroom remodeling can benefit from many of our recommendations.

Considering that you spend at least half of each day living in your house, you deserve the best living environment within your budget. I hope reading *New House/More House* has helped you come to the conclusion that you don't have to spend a fortune to make your residence something special. You will find that this industry has a lot of creative people who can build houses based on innovative ideas—and stay within a budget. You just have to take the time to find them. Your time is well spent if your search results in a house you enjoy coming home to every day.

Chapter Thirteen Recap

- Additions and remodelings have specific considerations beyond new house construction.

- Verify that real estate values in the area warrant additional investment in the property. Can the project cost be recovered if the house is sold?

- If you intend to stay for a long period of time, the enjoyment or utility of the project may outweigh investment concerns.

- You can determine for yourself the potential of an expansion project by visiting city hall to determine applicable zoning codes. If existing codes limit your ability to expand, consider a zoning variation for relief.

- You can also begin to assess the physical limitations of your site and house. Land contours, existing landscaping, and underground utilities can limit the potential of expansion projects.

- Older homes present additional problems for consideration. Utility services can be inadequate for additional space or amenities.

- Aging plumbing, electrical and structural framing systems may reveal unexpected problems only after construction has begun and walls have been opened. These surprises always cause unexpected budget problems.

- Additions require aesthetic design decisions regarding the exterior and interior of the house. You can choose either to continue the existing style or adopt a completely new look.

- Building upward or outward for an addition requires analysis of foundation structure and property restrictions.

- Families choosing to stay and live through an addition or remodeling project should anticipate stressful situations. Working with the project professionals, find ways to minimize disruption to daily routines.

- To save money on fees while determining if a project is feasible, hire a professional to conduct a short study to produce a conceptual design and construction estimate. This enables you to commit to a larger fee only if the project has merit and you wish to move ahead, and provides you with a good evaluation of the proposed project to make an informed decision.

Appendix

Figure 5-2, AIA Document B155

AIA Document B155

Standard Form of Agreement Between Owner and Architect for a Small Project

1993 SMALL PROJECTS EDITION

Because this document has important legal consequences, we encourage you to consult with an attorney before signing it. Some states mandate a cancellation period or require other specific disclosures, including warnings for home improvement contracts, when a document such as this will be used for Work on the Owner's personal residence. Your attorney should insert all language required by state or local law to be included in this Agreement. Such statements may be entered in the space provided below, or if required by law, above the signatures of the parties.

This **AGREEMENT** is made:
(Date)

BETWEEN the Owner:

and the Architect:

for the following Project:

The Owner and Architect agree as follows.

B155—1993 1

ARTICLE 1

ARCHITECT'S RESPONSIBILITIES

The Architect shall provide architectural services for the project, including normal structural, mechanical and electrical design services. Services shall be performed in a manner consistent with professional skill and care.

1.1 During the Design Phase, the Architect shall perform the following tasks:

   .1 describe the project requirements for the Owner's approval;

   .2 develop a design solution based on the approved project requirements;

   .3 upon the Owner's approval of the design solution, prepare Construction Documents indicating requirements for construction of the project;

   .4 assist the Owner in filing documents required for the approval of governmental authorities; and

   .5 assist the Owner in obtaining proposals and award contracts for construction.

1.2 During the Construction Phase, the Architect shall act as the Owner's representative and provide administration of the Contract between the Owner and Contractor. The extent of the Architect's authority and responsibility during construction is described in this Agreement and in AIA Document A205, General Conditions of the Contract for Construction of a Small Project. Unless otherwise agreed, the Architect's services during construction include visiting the site, reviewing and certifying payments, reviewing the Contractor's submittals, rejecting noncomforming Work, and interpreting the Contract Documents.

ARTICLE 2

OWNER'S RESPONSIBILITIES

The Owner shall provide full information about the objectives, schedule, constraints and existing conditions of the project, and shall establish a budget with reasonable contingencies that meets the project requirements. The Owner shall furnish surveying, geotechnical engineering and environmental testing services upon request by the Architect. The Owner shall employ a contractor to perform the construction Work and to provide cost-estimating services. The Owner shall furnish for the benefit of the project all legal, accounting and insurance counseling services.

ARTICLE 3

USE OF ARCHITECT'S DOCUMENTS

Documents prepared by the Architect are instruments of service for use solely with respect to this project. The Architect shall retain all common law, statutory and other reserved rights, including the copyright. The Owner shall not reuse or permit the reuse of the Architect's documents except by mutual agreement in writing.

ARTICLE 4

TERMINATION, SUSPENSION OR ABANDONMENT

In the event of termination, suspension or abandonment of the project, the Architect shall be equitably compensated for services performed. Failure of the Owner to make payments to the Architect in accordance with this Agreement shall be considered substantial nonperformance and is sufficient cause for the Architect to either suspend or terminate services. Either the Architect or the Owner may terminate this Agreement after giving no less than seven days' written notice if the other party substantially fails to perform in accordance with the terms of this Agreement.

ARTICLE 5

MISCELLANEOUS PROVISIONS

5.1 This Agreement shall be governed by the law of the location of the project.

5.2 Terms in this Agreement shall have the same meaning as those in AIA Document A205, General Conditions of the Contract for the Construction of a Small Project, current as of the date of this Agreement.

5.3 The Owner and Architect, respectively, bind themselves, their partners, successors, assigns and legal representatives to this Agreement. Neither party to this Agreement shall assign the contract as a whole without written consent of the other.

5.4 The Architect and Architect's consultants shall have no responsibility for the identification, discovery, presence, handling, removal or disposal of, or exposure of persons to, hazardous materials in any form at the project site.

B155—1993   2

Figure 5-2, AIA Document B155

ARTICLE 6

PAYMENTS AND COMPENSATION TO THE ARCHITECT

The Owner shall compensate the Architect as follows.

6.1 The Architect's Compensation shall be:
(Indicate method of compensation.)

of which an initial payment retainer of   dollars ($)
shall be paid upon execution of this Agreement and shall be credited to the final payment.

6.2 The Architect shall be reimbursed for expenses incurred in the interest of the project, plus an administrative fee of
                  percent (%).

(List reimbursable items.)

Sample

6.3 If through no fault of the Architect the services covered by this Agreement have not been completed within
(   ) months of the date hereof, compensation for the Architect's services beyond that
time shall be appropriately adjusted.

6.4 Payments are due and payable upon receipt of the Architect's invoice. Amounts unpaid
(   ) days after invoice date shall bear interest from the date payment is due at the rate of
         (   ), or in the absence thereof, at the legal rate prevailing at the principal place of business of the Architect.

(Usury laws and requirements under the Federal Truth in Lending Act, similar state and local consumer credit laws and other regulations at the Owner's and Architect's principal places of business, the location of the Project and elsewhere may affect the validity of this provision.)

6.5 Architectural services not covered by this Agreement include, among others, revisions due to changes in the scope, quality or budget. The Architect shall be paid additional fees for these services based on the Architect's hourly rates when the services are performed.

B155—1993   3

ARTICLE 7

OTHER PROVISIONS

(Insert descriptions of other services and modifications to the terms of this Agreement.)

Sample

This Agreement entered into as of the day and year first written above.
(If required by law, insert cancellation period, disclosures or other warning statements above the signatures.)

OWNER

_____
(Signature)

_____
(Printed name, title and address)

_____

_____

_____

ARCHITECT

_____
(Signature)

_____
(Printed name, title and address)

_____

_____

_____

Figure 5-3, AIA Document A105

AIA Document A105

Standard Form of Agreement Between Owner and Contractor for a Small Project

where the Basis of Payment is a STIPULATED SUM

1993 SMALL PROJECTS EDITION

Because this document has important legal consequences, we encourage you to consult with an attorney before signing it. Some states mandate a cancellation period or require other specific disclosures, including warnings for home improvement contracts, when a document such as this will be used for Work on the Owner's personal residence. Your attorney should insert all language required by state or local law to be included in this Agreement. Such statements may be entered in the space provided below, or if required by law, above the signatures of the parties.

Sample

This **AGREEMENT** is made:
(Date)

BETWEEN the Owner:

and the Contractor:

for the following Project:

The Architect is:

The Owner and Contractor agree as follows.

A105—1993   1

ARTICLE 1

THE CONTRACT DOCUMENTS

The Contractor shall complete the Work described in the Contract Documents for the project. The Contract Documents consist of:

   .1 this Agreement signed by the Owner and Contractor;

   .2 AIA Document A205, General Conditions of the Contract for Construction of a Small Project, current edition;

   .3 the Drawings and Specifications prepared by the Architect, dated
and enumerated as follows:

   Drawings:

   Specifications:

Sample

   .4 addenda prepared by the Architect as follows:

   .5 written change orders or orders for minor changes in the Work issued after execution of this Agreement; and

   .6 other documents, if any, identified as follows:

A105—1993   2

Figure 5-3, AIA Document A105

ARTICLE 2

DATE OF COMMENCEMENT AND SUBSTANTIAL COMPLETION DATE

The date of commencement shall be the date of this Agreement unless otherwise indicated below. The Contractor shall substantially complete the Work not later than   ,
subject to adjustment by Change Order.

(Insert the date or number of calendar days after the date of commencement.)

ARTICLE 3

CONTRACT SUM

3.1 Subject to additions and deductions by Change Order, the Contract Sum is:

3.2 For purposes of payment, the Contract Sum includes the following values related to portions of the Work:

Portion of Work   **Value**

Sample

3.3 The Contract Sum shall include all items and services necessary for the proper execution and completion of the Work.

A105—1993   3

ARTICLE 4

PAYMENT

4.1 Based on Contractor's Applications for Payment certified by the Architect, the Owner shall pay the Contractor as follows:
(Here insert payment procedures and provisions for retainage, if any.)

4.2 Payments due and unpaid under the Contract Documents shall bear interest from the date payment is due at the rate of
, or in the absence thereof, at the legal rate prevailing at the place of the Project.

(Usury laws and requirements under the Federal Truth in Lending Act, similar state and local consumer credit laws and other regulations at the Owner's and Contractor's principal places of business, the location of the Project and elsewhere may affect the validity of this provision.)

ARTICLE 5

INSURANCE

5.1 The Contractor shall provide Contractor's Liability and other Insurance as follows:
(Insert specific insurance required by the Owner.)

5.2 The Owner shall provide Owner's Liability and Owner's Property Insurance as follows:
(Insert specific insurance furnished by the Owner.)

5.3 The Contractor shall obtain an endorsement to its general liability insurance policy to cover the Contractor's obligations under Paragraph 3.12 of AIA Document A205, General Conditions of the Contract for Construction of Small Projects.

5.4 Certificates of insurance shall be provided by each party showing their respective coverages prior to commencement of the Work.

A105—1993   4

Figure 5-3, AIA Document A105

ARTICLE 6

OTHER TERMS AND CONDITIONS

(Insert any other terms or conditions below.)

This Agreement entered into as of the day and year first written above.
(If required by law, insert cancellation period, disclosures or other warning statements above the signatures.)

OWNER

_____

(Signature)

_____

(Printed name, title and address)

_____

_____

_____

CONTRACTOR

_____

(Signature)

_____

(Printed name, title and address)

_____

_____

_____

LICENSE NO. _____

JURISDICTION _____

AIA **CAUTION: You should sign an original AIA document which has this caution printed in red. An original assures that changes will not be obscured as may occur when documents are reproduced. See Instruction Sheet for Limited License for Reproduction of this document.**

AIA DOCUMENT A105 • OWNER-CONTRACTOR AGREEMENT—SMALL PROJECTS
EDITION • AIA® • ©1993 • THE AMERICAN INSTITUTE OF ARCHITECTS, 1735 NEW YORK
AVENUE, N.W., WASHINGTON, D.C. 20006-5292 • **WARNING: Unlicensed photocopying
violates U.S. copyright laws and will subject the violator to legal prosecution.**

A105—1993   5

Figure 5-4, AIA Document A205

AIA Document A205

General Conditions of the Contract for Construction of a Small Project

1993 SMALL PROJECTS EDITION

ARTICLE 1

GENERAL PROVISIONS

1.1 THE CONTRACT

The Contract represents the entire and integrated agreement between the parties and supersedes prior negotiations, representations or agreements, either written or oral. The Contract may be amended or modified only by a written modification.

1.2 THE WORK

The term "Work" means the construction and services required by the Contract Documents, and includes all other labor, materials, equipment and services provided by the Contractor to fulfill the Contractor's obligations.

1.3 INTENT

The intent of the Contract Documents is to include all items necessary for the proper execution and completion of the Work by the Contractor. The Contract Documents are complementary, and what is required by one shall be as binding as if required by all.

1.4 OWNERSHIP AND USE OF ARCHITECT'S DRAWINGS, SPECIFICATIONS AND OTHER DOCUMENTS

Documents prepared by the Architect are instruments of the Architect's service for use solely with respect to this project. The Architect shall retain all common law, statutory and other reserved rights, including the copyright. They are not to be used by the Contractor or any Subcontractor, Sub-subcontractor or material or equipment supplier for other projects or for additions to this project outside the scope of the Work without the specific written consent of the Owner and Architect.

ARTICLE 2

OWNER

2.1 INFORMATION AND SERVICES REQUIRED OF THE OWNER

2.1.1 If requested by the Contractor, the Owner shall furnish and pay for a survey and a legal description of the site.

2.1.2 Except for permits and fees which are the responsibility of the Contractor under the Contract Documents, the Owner shall obtain and pay for other necessary approvals, easements, assessments and charges.

2.2 OWNER'S RIGHT TO STOP THE WORK

If the Contractor fails to correct Work which is not in accordance with the Contract Documents, the Owner may direct the Contractor in writing to stop the Work until the correction is made.

2.3 OWNER'S RIGHT TO CARRY OUT THE WORK

If the Contractor defaults or neglects to carry out the Work in accordance with the Contract Documents and fails within a seven day period after receipt of written notice from the Owner to correct such default or neglect with diligence and promptness, the Owner may, without prejudice to other remedies, correct such deficiencies. In such case, a Change Order shall be issued deducting the cost of correction from payments due the Contractor.

2.4 OWNER'S RIGHT TO PERFORM CONSTRUCTION AND TO AWARD SEPARATE CONTRACTS

2.4.1 The Owner reserves the right to perform construction or operations related to the project with the Owner's own forces, and to award separate contracts in connection with other portions of the project.

2.4.2 The Contractor shall coordinate and cooperate with separate contractors employed by the Owner.

2.4.3 Costs caused by delays or by improperly timed activities or defective construction shall be borne by the party responsible therefor.

ARTICLE 3

CONTRACTOR

3.1 EXECUTION OF THE CONTRACT

Execution of the Contract by the Contractor is a representation that the Contractor has visited the site, become familiar with local conditions under which the Work is to be performed and correlated personal observations with requirements of the Contract Documents.

3.2 REVIEW OF CONTRACT DOCUMENTS AND FIELD CONDITIONS BY CONTRACTOR

The Contractor shall carefully study and compare the Contract Documents with each other and with information furnished by the Owner. Before commencing activities, the Contractor shall: (1) take field measurements and verify field conditions; (2) carefully compare this and other information known to

A205—1993   1

the Contractor with the Contract Documents; and (3) promptly report errors, inconsistencies or omissions discovered to the Architect.

3.3 SUPERVISION AND CONSTRUCTION PROCEDURES

3.3.1 The Contractor shall supervise and direct the Work, using the Contractor's best skill and attention. The Contractor shall be solely responsible for and have control over construction means, methods, techniques, sequences and procedures, and for coordinating all portions of the Work.

3.3.2 The Contractor, as soon as practicable after award of the Contract, shall furnish in writing to the Owner through the Architect the names of subcontractors or suppliers for each portion of the Work. The Architect will promptly reply to the Contractor in writing if the Owner or the Architect, after due investigation, has reasonable objection to the subcontractors or suppliers listed.

3.4 LABOR AND MATERIALS

3.4.1 Unless otherwise provided in the Contract Documents, the Contractor shall provide and pay for labor, materials, equipment, tools, utilities, transportation, and other facilities and services necessary for proper execution and completion of the Work.

3.4.2 The Contractor shall deliver, handle, store and install materials in accordance with manufacturers' instructions.

3.5 WARRANTY

The Contractor warrants to the Owner and Architect that: (1) materials and equipment furnished under the Contract will be new and of good quality unless otherwise required or permitted by the Contract Documents; (2) the Work will be free from defects not inherent in the quality required or permitted; and (3) the Work will conform to the requirements of the Contract Documents.

3.6 TAXES

The Contractor shall pay sales, consumer, use and similar taxes that are legally required when the Contract is executed.

3.7 PERMITS, FEES AND NOTICES

3.7.1 The Contractor shall obtain and pay for the building permit and other permits and governmental fees, licenses and inspections necessary for proper execution and completion of the Work.

3.7.2 The Contractor shall comply with and give notices required by agencies having jurisdiction over the Work. If the Contractor performs Work knowing it to be contrary to laws, statutes, ordinances, building codes, and rules and regulations without notice to the Architect and Owner, the Contractor shall assume full responsibility for such Work and shall bear the attributable costs. The Contractor shall promptly notify the Architect in writing of any known inconsistencies in the Contract Documents with such governmental laws, rules and regulations.

3.8 SUBMITTALS

The Contractor shall promptly review, approve in writing and submit to the Architect Shop Drawings, Product Data, Samples and similar submittals required by the Contract Documents. Shop Drawings, Product Data, Samples and similar submittals are not Contract Documents.

3.9 USE OF SITE

The Contractor shall confine operations at the site to areas permitted by law, ordinances, permits, the Contract Documents and the Owner.

3.10 CUTTING AND PATCHING

The Contractor shall be responsible for cutting, fitting or patching required to complete the Work or to make its parts fit together properly.

3.11 CLEANING UP

The Contractor shall keep the premises and surrounding area free from accumulation of debris and trash related to the Work.

3.12 INDEMNIFICATION

To the fullest extent permitted by law, the Contractor shall indemnify and hold harmless the Owner, Architect, Architect's consultants and agents and employees of any of them from and against claims, damages, losses and expenses, including but not limited to attorneys' fees, arising out of or resulting from performance of the Work, provided that such claim, damage, loss or expense is attributable to bodily injury, sickness, disease or death, or to injury to or destruction of tangible property (other than the Work itself) including loss of use resulting therefrom, but only to the extent caused in whole or in part by negligent acts or omissions of the Contractor, a Subcontractor, anyone directly or indirectly employed by them or anyone for whose acts they may be liable, regardless of whether or not such claim, damage, loss or expense is caused in part by a party indemnified hereunder.

ARTICLE 4

ARCHITECT'S ADMINISTRATION OF THE CONTRACT

4.1 The Architect will provide administration of the Contract as described in the Contract Documents. The Architect will have authority to act on behalf of the Owner only to the extent provided in the Contract Documents.

4.2 The Architect will visit the site at intervals appropriate to the stage of construction to become generally familiar with the progress and quality of the Work.

4.3 The Architect will not have control over or charge of and will not be responsible for construction means, methods, techniques, sequences or procedures, or for safety precautions and programs in connection with the Work, since these are solely the Contractor's responsibility. The Architect will not be responsible for the Contractor's failure to carry out the Work in accordance with the Contract Documents.

4.4 Based on the Architect's observations and evaluations of the Contractor's Applications for Payment, the Architect will review and certify the amounts due the Contractor.

4.5 The Architect will have authority to reject Work that does not conform to the Contract Documents.

4.6 The Architect will promptly review and approve or take appropriate action upon Contractor's submittals such as Shop Drawings, Product Data and Samples, but only for the limited purpose of checking for conformance with information given and the design concept expressed in the Contract Documents.

4.7 The Architect will promptly interpret and decide matters concerning performance under and requirements of the

Figure 5-4, AIA Document A205

Contract Documents on written request of either the Owner or Contractor.

4.8 Interpretations and decisions of the Architect will be consistent with the intent of and reasonably inferable from the Contract Documents and will be in writing or in the form of drawings. When making such interpretations and decisions, the Architect will endeavor to secure faithful performance by both Owner and Contractor, will not show partiality to either and will not be liable for results of interpretations or decisions so rendered in good faith.

4.9 The Architect's duties, responsibilities and limits of authority as described in the Contract Documents will not be changed without written consent of the Owner, Contractor and Architect. Consent shall not be unreasonably withheld.

ARTICLE 5

CHANGES IN THE WORK

5.1 After execution of the Contract, changes in the Work may be accomplished by Change Order or by order for a minor change in the Work. The Owner, without invalidating the Contract, may order changes in the Work within the general scope of the Contract consisting of additions, deletions or other revisions, the Contract Sum and Contract Time being adjusted accordingly.

5.2 A Change Order shall be a written order to the Contractor signed by the Owner and Architect to change the Work, Contract Sum or Contract Time.

5.3 The Architect will have authority to order minor changes in the Work not involving changes in the Contract Sum or the Contract Time and not inconsistent with the intent of the Contract Documents. Such changes shall be written orders and shall be binding on the Owner and Contractor. The Contractor shall carry out such written orders promptly.

5.4 If concealed or unknown physical conditions are encountered at the site that differ materially from those indicated in the Contract Documents or from those conditions ordinarily found to exist, the Contract Sum and Contract Time shall be subject to equitable adjustment.

ARTICLE 6

TIME

6.1 Time limits stated in the Contract Documents are of the essence of the Contract.

6.2 If the Contractor is delayed at any time in progress of the Work by changes ordered in the Work, or by labor disputes, fire, unusual delay in deliveries, unavoidable casualties or other causes beyond the Contractor's control, the Contract Time shall be extended by Change Order for such reasonable time as the Architect may determine.

ARTICLE 7

PAYMENTS AND COMPLETION

7.1 CONTRACT SUM

The Contract Sum stated in the Agreement, including authorized adjustments, is the total amount payable by the Owner to

the Contractor for performance of the Work under the Contract Documents.

7.2 APPLICATIONS FOR PAYMENT

7.2.1 At least ten days before the date established for each progress payment, the Contractor shall submit to the Architect an itemized Application for Payment for operations completed in accordance with the values stated in the Agreement. Such application shall be supported by such data substantiating the Contractor's right to payment as the Owner or Architect may reasonably require and reflecting retainage if provided for elsewhere in the Contract Documents.

7.2.2 The Contractor warrants that title to all Work covered by an Application for Payment will pass to the Owner no later than the time of payment. The Contractor further warrants that upon submittal of an Application for Payment, all Work for which Certificates for Payment have been previously issued and payments received from the Owner shall, to the best of the Contractor's knowledge, information and belief, be free and clear of liens, claims, security interests or other encumbrances adverse to the Owner's interests.

7.3 CERTIFICATES FOR PAYMENT

The Architect will, within seven days after receipt of the Contractor's Application for Payment, either issue to the Owner a Certificate for Payment, with a copy to the Contractor, for such amount as the Architect determines is properly due, or notify the Contractor and Owner in writing of the Architect's reasons for withholding certification in whole or in part.

7.4 PROGRESS PAYMENTS

7.4.1 After the Architect has issued a Certificate for Payment, the Owner shall make payment in the manner provided in the Contract Documents.

7.4.2 The Contractor shall promptly pay each Subcontractor and material supplier, upon receipt of payment from the Owner, out of the amount paid to the Contractor on account of such entities' portion of the Work.

7.4.3 Neither the Owner nor the Architect shall have responsibility for the payment of money to a Subcontractor or material supplier.

7.4.4 A Certificate for Payment, a progress payment, or partial or entire use or occupancy of the project by the Owner shall not constitute acceptance of Work not in accordance with the requirements of the Contract Documents.

7.5 SUBSTANTIAL COMPLETION

7.5.1 Substantial Completion is the stage in the progress of the Work when the Work or designated portion thereof is sufficiently complete in accordance with the Contract Documents so that the Owner can occupy or utilize the Work for its intended use.

7.5.2 When the Work or designated portion thereof is substantially complete, the Architect will prepare a Certificate of Substantial Completion which shall establish the date of Substantial Completion, shall establish the responsibilities of the Owner and Contractor, and shall fix the time within which the Contractor shall finish all items on the list accompanying the Certificate. Warranties required by the Contract Documents shall commence on the date of Substantial Completion of the Work or designated portion thereof unless otherwise provided in the Certificate of Substantial Completion.

7.6 FINAL COMPLETION AND FINAL PAYMENT

7.6.1 Upon receipt of a final Application for Payment, the Architect will inspect the Work. When the Architect finds the Work acceptable and the Contract fully performed, the Architect will promptly issue a final Certificate for Payment.

7.6.2 Final payment shall not become due until the Contractor submits to the Architect releases and waivers of liens, and data establishing payment or satisfaction of obligations, such as receipts, claims, security interests or encumbrances arising out of the Contract.

7.6.3 Acceptance of final payment by the Contractor, a Subcontractor or material supplier shall constitute a waiver of claims by that payee except those previously made in writing and identified by that payee as unsettled at the time of final Application for Payment.

ARTICLE 8

PROTECTION OF PERSONS AND PROPERTY

8.1 SAFETY PRECAUTIONS AND PROGRAMS

The Contractor shall be responsible for initiating, maintaining and supervising all safety precautions and programs, including all those required by law in connection with performance of the Contract. The Contractor shall promptly remedy damage and loss to property caused in whole or in part by the Contractor, or by anyone for whose acts the Contractor may be liable.

ARTICLE 9

CORRECTION OF WORK

9.1 The Contractor shall promptly correct Work rejected by the Architect as failing to conform to the requirements of the Contract Documents. The Contractor shall bear the cost of correcting such rejected Work.

9.2 In addition to the Contractor's other obligations including warranties under the Contract, the Contractor shall, for a period of one year after Substantial Completion, correct work not conforming to the requirements of the Contract Documents.

9.3 If the Contractor fails to correct nonconforming Work within a reasonable time, the Owner may correct it and the Contractor shall reimburse the Owner for the cost of correction.

ARTICLE 10

MISCELLANEOUS PROVISIONS

10.1 ASSIGNMENT OF CONTRACT

Neither party to the Contract shall assign the Contract as a whole without written consent of the other.

10.2 TESTS AND INSPECTIONS

10.2.1 Tests, inspections and approvals of portions of the Work required by the Contract Documents or by laws, ordi-

nances, rules, regulations or orders of public authorities having jurisdiction shall be made at an appropriate time.

10.2.2 If the Architect requires additional testing, the Contractor shall perform these tests.

10.2.3 The Owner shall pay for tests except for testing Work found to be defective for which the Contractor shall pay.

10.3 GOVERNING LAW

The Contract shall be governed by the law of the place where the project is located.

ARTICLE 11

TERMINATION OF THE CONTRACT

11.1 TERMINATION BY THE CONTRACTOR

If the Owner fails to make payment when due or substantially breaches any other obligation of this Contract, following seven days' written notice to the Owner, the Contractor may terminate the Contract and recover from the Owner payment for Work executed and for proven loss with respect to materials, equipment, tools, construction equipment and machinery, including reasonable overhead, profit and damages.

11.2 TERMINATION BY THE OWNER

11.2.1 The Owner may terminate the Contract if the Contractor:

  .1 persistently or repeatedly refuses or fails to supply enough properly skilled workers or proper materials;

  .2 fails to make payment to Subcontractors for materials or labor in accordance with the respective agreements between the Contractor and the Subcontractors;

  .3 persistently disregards laws, ordinances, or rules, regulations or orders of a public authority having jurisdiction; or

  .4 is otherwise guilty of substantial breach of a provision of the Contract Documents.

11.2.2 When any of the above reasons exist, the Owner, after consultation with the Architect, may without prejudice to any other rights or remedies of the Owner and after giving the Contractor and the Contractor's surety, if any, seven days' written notice, terminate employment of the Contractor and may:

  .1 take possession of the site and of all materials thereon owned by the Contractor;

  .2 finish the Work by whatever reasonable method the Owner may deem expedient.

11.2.3 When the Owner terminates the Contract for one of the reasons stated in Subparagraph 11.2.1, the Contractor shall not be entitled to receive further payment until the Work is finished.

11.2.4 If the unpaid balance of the Contract Sum exceeds costs of finishing the Work, such excess shall be paid to the Contractor. If such costs exceed the unpaid balance, the Contractor shall pay the difference to the Owner. This obligation for payment shall survive termination of the Contract.

4   A205—1993

Figure 5-5, AIA Document A191

AIA Document A191

Standard Form of Agreements Between Owner and Design/Builder

THIS DOCUMENT HAS IMPORTANT LEGAL CONSEQUENCES; CONSULTATION WITH AN ATTORNEY IS ENCOURAGED WITH RESPECT TO ITS USE, COMPLETION OR MODIFICATION.

1996 EDITION

TABLE OF ARTICLES

PART 1 AGREEMENT

PART 2 AGREEMENT

AIA Document A191

Standard Form of Agreement Between Owner and Design/Builder

THIS DOCUMENT HAS IMPORTANT LEGAL CONSEQUENCES; CONSULTATION WITH AN ATTORNEY IS ENCOURAGED WITH RESPECT TO ITS USE, COMPLETION OR MODIFICATION.

This document comprises two separate Agreements: Part 1 Agreement and Part 2 Agreement. Before executing the Part 1 Agreement, the parties should reach substantial agreement on the Part 2 Agreement. To the extent referenced in these Agreements, subordinate parallel agreements to A191 consist of AIA Document A491, Standard Form of Agreements Between Design/Builder and Contractor, and AIA Document B901, Standard Form of Agreements Between Design/Builder and Architect.

PART 1 AGREEMENT

1996 EDITION

AGREEMENT

made as of the .... day of .... in the year of
(In words, indicate day, month and year.)

BETWEEN the Owner:
(Name and address)

and the Design/Builder:
(Name and address)

**A191—1996
Part 1—Page 1**

Figure 5-5, AIA Document A191

For the following Project:
(Include Project name, location and a summary description.)

The architectural services described in Article 1 will be provided by the following person or entity who is lawfully licensed
to practice architecture:
(Name and address)   *(Registration Number)*   *(Relationship to Design/Builder)*

Normal structural, mechanical and electrical engineering services will be provided contractually through the Architect except
as indicated below:
(Name, address and discipline)   *(Registration Number)*   *(Relationship to Design/Builder)*

The Owner and the Design/Builder agree as set forth below.

**A191—1996
Part 1—Page 2**

TERMS AND CONDITIONS—PART 1 AGREEMENT

ARTICLE 1
DESIGN/BUILDER

1.1 SERVICES

1.1.1 Preliminary design, budget, and schedule comprise the services required to accomplish the preparation and submission of the Design/Builder's Proposal as well as the preparation and submission of any modifications to the Proposal prior to execution of the Part 2 Agreement.

1.2 RESPONSIBILITIES

1.2.1 Design services required by this Part 1 Agreement shall be performed by qualified architects and other design professionals. The contractual obligations of such professional persons or entities are undertaken and performed in the interest of the Design/Builder.

1.2.2 The agreements between the Design/Builder and the persons or entities identified in this Part 1 Agreement, and any subsequent modifications, shall be in writing. These agreements, including financial arrangements with respect to this Project, shall be promptly and fully disclosed to the Owner upon request.

1.2.3 Construction budgets shall be prepared by qualified professionals, cost estimators or contractors retained by and acting in the interest of the Design/Builder.

1.2.4 The Design/Builder shall be responsible to the Owner for acts and omissions of the Design/Builder's employees, subcontractors and their agents and employees, and other persons, including the Architect and other design professionals, performing any portion of the Design/Builder's obligations under this Part 1 Agreement.

1.2.5 If the Design/Builder believes or is advised by the Architect or by another design professional retained to provide services on the Project that implementation of any instruction received from the Owner would cause a violation of any applicable law, the Design/Builder shall notify the Owner in writing. Neither the Design/Builder nor the Architect shall be obligated to perform any act which either believes will violate any applicable law.

1.2.6 Nothing contained in this Part 1 Agreement shall create a contractual relationship between the Owner and any person or entity other than the Design/Builder.

1.3 BASIC SERVICES

1.3.1 The Design/Builder shall provide a preliminary evaluation of the Owner's program and project budget requirements, each in terms of the other.

1.3.2 The Design/Builder shall visit the site, become familiar with the local conditions, and correlate observable conditions with the requirements of the Owner's program, schedule and budget.

1.3.3 The Design/Builder shall review laws applicable to design and construction of the Project, correlate such laws with the Owner's program requirements, and advise the Owner if any program requirement may cause a violation of such laws. Necessary changes to the Owner's program shall be accomplished by appropriate written modification or disclosed as described in Paragraph 1.3.5.

1.3.4 The Design/Builder shall review with the Owner alternative approaches to design and construction of the Project.

1.3.5 The Design/Builder shall submit to the Owner a Proposal, including the completed Preliminary Design Documents, a statement of the proposed contract sum, and a proposed schedule for completion of the Project. Preliminary Design Documents shall consist of preliminary design drawings, outline specifications or other documents sufficient to establish the size, quality and character of the entire Project, its architectural, structural, mechanical and electrical systems, and the materials and such other elements of the Project as may be appropriate. Deviations from the Owner's program shall be disclosed in the Proposal. If the Proposal is accepted by the Owner, the parties shall then execute the Part 2 Agreement. A modification to the Proposal before execution of the Part 2 Agreement shall be recorded in writing as an addendum and shall be identified in the Contract Documents of the Part 2 Agreement.

1.4 ADDITIONAL SERVICES

1.4.1 The Additional Services described under this Paragraph 1.4 shall be provided by the Design/Builder and paid for by the Owner if authorized or confirmed in writing by the Owner.

1.4.2 Making revisions in the Preliminary Design Documents, budget or other documents when such revisions are:

.1 inconsistent with approvals or instructions previously given by the Owner, including revisions made necessary by adjustments in the Owner's program or Project budget;

.2 required by the enactment or revision of codes, laws or regulations subsequent to the preparation of such documents; or

.3 due to changes required as a result of the Owner's failure to render decisions in a timely manner.

1.4.3 Providing more extensive programmatic criteria than that furnished by the Owner as described in Paragraph 2.1. When authorized, the Design/Builder shall provide professional services to assist the Owner in the preparation of the program. Programming services may consist of:

.1 consulting with the Owner and other persons or entities not designated in this Part 1 Agreement to define the program requirements of the Project and to review the understanding of such requirements with the Owner;

Figure 5-5, AIA Document A191

.2 documentation of the applicable requirements necessary for the various Project functions or operations;

.3 providing a review and analysis of the functional and organizational relationships, requirements, and objectives for the Project;

.4 setting forth a written program of requirements for the Owner's approval which summarizes the Owner's objectives, schedule, constraints, and criteria.

1.4.4 Providing financial feasibility or other special studies.

1.4.5 Providing planning surveys, site evaluations or comparative studies of prospective sites.

1.4.6 Providing special surveys, environmental studies, and submissions required for approvals of governmental authorities or others having jurisdiction over the Project.

1.4.7 Providing services relative to future facilities, systems and equipment.

1.4.8 Providing services at the Owner's specific request to perform detailed investigations of existing conditions or facilities or to make measured drawings thereof.

1.4.9 Providing services at the Owner's specific request to verify the accuracy of drawings or other information furnished by the Owner.

1.4.10 Coordinating services in connection with the work of separate persons or entities retained by the Owner, subsequent to the execution of this Part 1 Agreement.

1.4.11 Providing analyses of owning and operating costs.

1.4.12 Providing interior design and other similar services required for or in connection with the selection, procurement or installation of furniture, furnishings and related equipment.

1.4.13 Providing services for planning tenant or rental spaces.

1.4.14 Making investigations, inventories of materials or equipment, or valuations and detailed appraisals of existing facilities.

ARTICLE 2

OWNER

2.1 RESPONSIBILITIES

2.1.1 The Owner shall provide full information in a timely manner regarding requirements for the Project, including a written program which shall set forth the Owner's objectives, schedule, constraints and criteria.

2.1.2 The Owner shall establish and update an overall budget for the Project, including reasonable contingencies. This budget shall not constitute the contract sum.

2.1.3 The Owner shall designate a representative authorized to act on the Owner's behalf with respect to the

Project. The Owner or such authorized representative shall render decisions in a timely manner pertaining to documents submitted by the Design/Builder in order to avoid unreasonable delay in the orderly and sequential progress of the Design/Builder's services. The Owner may obtain independent review of the documents by a separate architect, engineer, contractor or cost estimator under contract to or employed by the Owner. Such independent review shall be undertaken at the Owner's expense in a timely manner and shall not delay the orderly progress of the Design/Builder's services.

2.1.4 The Owner shall furnish surveys describing physical characteristics, legal limitations and utility locations for the site of the Project, and a written legal description of the site. The surveys and legal information shall include, as applicable, grades and lines of streets, alleys, pavements, and adjoining property and structures; adjacent drainage; rights-of-way, restrictions, easements, encroachments, zoning, deed restrictions, boundaries and contours of the site; locations, dimensions and necessary data pertaining to existing buildings, other improvements and trees; and information concerning available utility services and lines, both public and private, above and below grade, including inverts and depths. All the information on the survey shall be referenced to a Project benchmark.

2.1.5 The Owner shall furnish the services of geotechnical engineers when such services are stipulated in this Part 1 Agreement, or deemed reasonably necessary by the Design/Builder. Such services may include but are not limited to test borings, test pits, determinations of soil bearing values, percolation tests, evaluations of hazardous materials, ground corrosion and resistivity tests, and necessary operations for anticipating subsoil conditions. The services of geotechnical engineer(s) or other consultants shall include preparation and submission of all appropriate reports and professional recommendations.

2.1.6 The Owner shall disclose, to the extent known to the Owner, the results and reports of prior tests, inspections or investigations conducted for the Project involving: structural or mechanical systems; chemical, air and water pollution; hazardous materials; or other environmental and subsurface conditions. The Owner shall disclose all information known to the Owner regarding the presence of pollutants at the Project's site.

2.1.7 The Owner shall furnish all legal, accounting and insurance counseling services as may be necessary at any time for the Project, including such auditing services as the Owner may require to verify the Design/Builder's Applications for Payment.

2.1.8 The Owner shall promptly obtain easements, zoning variances and legal authorizations regarding site utilization where essential to the execution of the Owner's program.

2.1.9 Those services, information, surveys and reports required by Paragraphs 2.1.4 through 2.1.8 which are within the Owner's control shall be furnished at the Owner's expense, and the Design/Builder shall be entitled to rely upon the accuracy and completeness thereof, except to the extent the Owner advises the Design/Builder to the contrary in writing.

2.1.10 If the Owner requires the Design/Builder to maintain any special insurance coverage, policy, amendment, or rider, the Owner shall pay the additional cost thereof except as otherwise stipulated in this Part 1 Agreement.

2.1.11 The Owner shall communicate with persons or entities employed or retained by the Design/Builder through the Design/Builder, unless otherwise directed by the Design/Builder.

ARTICLE 3

OWNERSHIP AND USE OF DOCUMENTS AND ELECTRONIC DATA

3.1 Drawings, specifications, and other documents and electronic data furnished by the Design/Builder are instruments of service. The Design/Builder's Architect and other providers of professional services shall retain all common law, statutory and other reserved rights, including copyright in those instruments of service furnished by them. Drawings, specifications, and other documents and electronic data are furnished for use solely with respect to this Part 1 Agreement. The Owner shall be permitted to retain copies, including reproducible copies, of the drawings, specifications, and other documents and electronic data furnished by the Design/Builder for information and reference in connection with the Project except as provided in Paragraphs 3.2 and 3.3.

3.2 If the Part 2 Agreement is not executed, the Owner shall not use the drawings, specifications, and other documents and electronic data furnished by the Design/Builder without the written permission of the Design/Builder. Drawings, specifications, and other documents and electronic data shall not be used by the Owner or others on other projects, for additions to this Project or for completion of this Project by others, except by agreement in writing and with appropriate compensation to the Design/Builder, unless the Design/Builder is adjudged to be in default under this Part 1 Agreement or under any other subsequently executed agreement, or by agreement in writing.

3.3 If the Design/Builder defaults in the Design/Builder's obligations to the Owner, the Architect shall grant a license to the Owner to use the drawings, specifications, and other documents and electronic data furnished by the Architect to the Design/Builder for the completion of the Project, conditioned upon the Owner's execution of an agreement to cure the Design/Builder's default in payment to the Architect for services previously performed and to indemnify the Architect with regard to claims arising from such reuse without the Architect's professional involvement.

3.4 Submission or distribution of the Design/Builder's documents to meet official regulatory requirements or for similar purposes in connection with the Project is not to be construed as publication in derogation of the rights reserved in Paragraph 3.1.

ARTICLE 4

TIME

4.1 Upon the request of the Owner, the Design/Builder shall prepare a schedule for the performance of the Basic and Additional Services which shall not exceed the time limits contained in Paragraph 10.1 and shall include allowances for periods of time required for the Owner's review and for approval of submissions by authorities having jurisdiction over the Project.

4.2 If the Design/Builder is delayed in the performance of services under this Part 1 Agreement through no fault of the Design/Builder, any applicable schedule shall be equitably adjusted.

ARTICLE 5

PAYMENTS

5.1 The initial payment provided in Article 9 shall be made upon execution of this Part 1 Agreement and credited to the Owner's account as provided in Subparagraph 9.1.2.

5.2 Subsequent payments for Basic Services, Additional Services, and Reimbursable Expenses provided for in this Part 1 Agreement shall be made monthly on the basis set forth in Article 9.

5.3 Within ten (10) days of the Owner's receipt of a properly submitted and correct Application for Payment, the Owner shall make payment to the Design/Builder.

5.4 Payments due the Design/Builder under this Part 1 Agreement which are not paid when due shall bear interest from the date due at the rate specified in Paragraph 9.5, or in the absence of a specified rate, at the legal rate prevailing where the Project is located.

ARTICLE 6

DISPUTE RESOLUTION— MEDIATION AND ARBITRATION

6.1 Claims, disputes or other matters in question between the parties to this Part 1 Agreement arising out of or relating to this Part 1 Agreement or breach thereof shall be subject to and decided by mediation or arbitration. Such mediation or arbitration shall be conducted in accordance with the Construction Industry Mediation or Arbitration Rules of the American Arbitration Association currently in effect.

6.2 In addition to and prior to arbitration, the parties shall endeavor to settle disputes by mediation. Demand for mediation shall be filed in writing with the other party to this Part 1 Agreement and with the American Arbitration Association. A demand for mediation shall be made within a reasonable time after the claim, dispute or other matter in question has arisen. In no event shall the demand for mediation be made after the date when

Figure 5-5, AIA Document A191

institution of legal or equitable proceedings based on such claim, dispute or other matter in question would be barred by the applicable statute of repose or limitations.

6.3 Demand for arbitration shall be filed in writing with the other party to this Part 1 Agreement and with the American Arbitration Association. A demand for arbitration shall be made within a reasonable time after the claim, dispute or other matter in question has arisen. In no event shall the demand for arbitration be made after the date when institution of legal or equitable proceedings based on such claim, dispute or other matter in question would be barred by the applicable statutes of repose or limitations.

6.4 An arbitration pursuant to this Paragraph may be joined with an arbitration involving common issues of law or fact between the Design/Builder and any person or entity with whom the Design/Builder has a contractual obligation to arbitrate disputes. No other arbitration arising out of or relating to this Part 1 Agreement shall include, by consolidation, joinder or in any other manner, an additional person or entity not a party to this Part 1 Agreement or not a party to an agreement with the Design/Builder, except by written consent containing a specific reference to this Part 1 Agreement signed by the Owner, the Design/Builder and all other persons or entities sought to be joined. Consent to arbitration involving an additional person or entity shall not constitute consent to arbitration of any claim, dispute or other matter in question not described in the written consent or with a person or entity not named or described therein. The foregoing agreement to arbitrate and other agreements to arbitrate with an additional person or entity duly consented to by the parties to this Part 1 Agreement shall be specifically enforceable in accordance with applicable law in any court having jurisdiction thereof.

6.5 The award rendered by the arbitrator or arbitrators shall be final, and judgment may be entered upon it in accordance with applicable law in any court having jurisdiction thereof.

ARTICLE 7
MISCELLANEOUS PROVISIONS

7.1 Unless otherwise provided, this Part 1 Agreement shall be governed by the law of the place where the Project is located.

7.2 The Owner and the Design/Builder, respectively, bind themselves, their partners, successors, assigns and legal representatives to the other party to this Part 1 Agreement and to the partners, successors and assigns of such other party with respect to all covenants of this Part 1 Agreement. Neither the Owner nor the Design/Builder shall assign this Part 1 Agreement without the written consent of the other.

7.3 Unless otherwise provided, neither the design for nor the cost of remediation of hazardous materials shall be the responsibility of the Design/Builder.

7.4 This Part 1 Agreement represents the entire and integrated agreement between the Owner and the Design/Builder and supersedes all prior negotiations, representations or agreements, either written or oral. This Part 1 Agreement may be amended only by written instrument signed by both the Owner and the Design/Builder.

7.5 Prior to the termination of the services of the Architect or any other design professional designated in this Part 1 Agreement, the Design/Builder shall identify to the Owner in writing another architect or design professional with respect to whom the Owner has no reasonable objection, who will provide the services originally to have been provided by the Architect or other design professional whose services are being terminated.

ARTICLE 8
TERMINATION OF THE AGREEMENT

8.1 This Part 1 Agreement may be terminated by either party upon seven (7) days' written notice should the other party fail to perform substantially in accordance with its terms through no fault of the party initiating the termination.

8.2 This Part 1 Agreement may be terminated by the Owner without cause upon at least seven (7) days' written notice to the Design/Builder.

8.3 In the event of termination not the fault of the Design/Builder, the Design/Builder shall be compensated for services performed to the termination date, together with Reimbursable Expenses then due and Termination Expenses. Termination Expenses are expenses directly attributable to termination, including a reasonable amount for overhead and profit, for which the Design/Builder is not otherwise compensated under this Part 1 Agreement.

ARTICLE 9

BASIS OF COMPENSATION

The Owner shall compensate the Design/Builder in accordance with Article 5, Payments, and the other provisions of this Part 1 Agreement as described below.

9.1 **COMPENSATION FOR BASIC SERVICES**

9.1.1 FOR BASIC SERVICES, compensation shall be as follows:

9.1.2 AN INITIAL PAYMENT of   Dollars ($) shall be made upon execution of this Part 1 Agreement and credited to the Owner's account as follows:

9.1.3 SUBSEQUENT PAYMENTS shall be as follows:

9.2 **COMPENSATION FOR ADDITIONAL SERVICES**

9.2.1 FOR ADDITIONAL SERVICES, compensation shall be as follows:

9.3 **REIMBURSABLE EXPENSES**

9.3.1 Reimbursable Expenses are in addition to Compensation for Basic and Additional Services, and include actual expenditures made by the Design/Builder and the Design/Builder's employees and contractors in the interest of the Project, as follows:

9.3.2 FOR REIMBURSABLE EXPENSES, compensation shall be a multiple of (   ) times the amounts expended.

Figure 5-5, AIA Document A191

9.4 DIRECT PERSONNEL EXPENSE is defined as the direct salaries of personnel engaged on the Project, and the portion of the cost of their mandatory and customary contributions and benefits related thereto, such as employment taxes and other statutory employee benefits, insurance, sick leave, holidays, vacations, pensions and similar contributions and benefits.

9.5 INTEREST PAYMENTS

9.5.1 The rate of interest for past due payments shall be as follows:

(Usury laws and requirements under the Federal Truth in Lending Act, similar state and local consumer credit laws and other regulations at the Owner's and Design/Builder's principal places of business, at the location of the Project and elsewhere may affect the validity of this provision. Specific legal advice should be obtained with respect to deletion, modification or other requirements, such as written disclosures or waivers.)

9.6 IF THE SCOPE of the Project is changed materially, the amount of compensation shall be equitably adjusted.

9.7 The compensation set forth in this Part 1 Agreement shall be equitably adjusted if through no fault of the Design/Builder the services have not been completed within   (   ) months of the date of this Part 1 Agreement.

ARTICLE 10
OTHER CONDITIONS AND SERVICES

10.1 The Basic Services to be performed shall be commenced on   and, subject to authorized adjustments and to delays not caused by the Design/Builder, shall be completed in (   ) calendar days. The Design/Builder's Basic Services consist of those described in Paragraph 1.3 as part of Basic Services, and include normal professional engineering and preliminary design services, unless otherwise indicated.

10.2 Services beyond those described in Paragraph 1.4 are as follows:
(Insert descriptions of other services, identify Additional Services included within Basic Compensation and modifications to the payment and compensation terms included in this Agreement.)

10.3 The Owner's preliminary program, budget and other documents, if any, are enumerated as follows:

Title   Date

Sample

This Agreement entered into as of the day and year first written above.

OWNER   DESIGN/BUILDER

_____   _____
(Signature)   *(Signature)*

_____   _____
(Printed name and title)   *(Printed name and title)*

 CAUTION: You should sign an original AIA document which has this caution printed in red. An original assures that changes will not be obscured as may occur when documents are reproduced.

Figure 5-5, AIA Document A191

AIA Document A191

Standard Form of Agreement Between Owner and Design/Builder

THIS DOCUMENT HAS IMPORTANT LEGAL CONSEQUENCES; CONSULTATION WITH AN ATTORNEY IS ENCOURAGED WITH RESPECT TO ITS USE, COMPLETION OR MODIFICATION.

This document comprises two separate Agreements: Part 1 Agreement and Part 2 Agreement. To the extent referenced in these Agreements, subordinate parallel agreements to A191 consist of AIA Document A491, Standard Form of Agreements Between Design/Builder and Contractor, and AIA Document B901, Standard Form of Agreements Between Design/Builder and Architect.

PART 2 AGREEMENT

1996 EDITION

AGREEMENT

made as of the   day of   in the year of
(In words, indicate day, month and year.)

Sample

BETWEEN the Owner:
(Name and address)

and the Design/Builder:
(Name and address)

A191—1996
Part 2—Page 1

For the following Project:
(Include Project name, location and a summary description.)

The architectural services described in Article 3 will be provided by the following person or entity who is lawfully licensed to practice architecture:
(Name and address)   *(Registration Number)*   *(Relationship to Design/Builder)*

Normal structural, mechanical and electrical engineering services will be provided contractually through the Architect except as indicated below:
(Name, address and discipline)   *(Registration Number)*   *(Relationship to Design/Builder)*

The Owner and the Design/Builder agree as set forth below.

A191—1996
Part 2—Page 2

Figure 5-5, AIA Document A191

TERMS AND CONDITIONS—PART 2 AGREEMENT

ARTICLE 1
GENERAL PROVISIONS

1.1 BASIC DEFINITIONS

1.1.1 The Contract Documents consist of the Part 1 Agreement to the extent not modified by this Part 2 Agreement, this Part 2 Agreement, the Design/Builder's Proposal and written addenda to the Proposal identified in Article 14, the Construction Documents approved by the Owner in accordance with Subparagraph 3.2.3 and Modifications issued after execution of this Part 2 Agreement. A Modification is a Change Order or a written amendment to this Part 2 Agreement signed by both parties, or a Construction Change Directive issued by the Owner in accordance with Paragraph 8.3.

1.1.2 The term "Work" means the construction and services provided by the Design/Builder to fulfill the Design/Builder's obligations.

1.2 EXECUTION, CORRELATION AND INTENT

1.2.1 It is the intent of the Owner and the Design/Builder that the Contract Documents include all items necessary for proper execution and completion of the Work. The Contract Documents are complementary, and what is required by one shall be as binding as if required by all; performance by the Design/Builder shall be required only to the extent consistent with and reasonably inferable from the Contract Documents as being necessary to produce the intended results. Words that have well-known technical or construction industry meanings are used in the Contract Documents in accordance with such recognized meanings.

1.2.2 If the Design/Builder believes or is advised by the Architect or by another design professional retained to provide services on the Project that implementation of any instruction received from the Owner would cause a violation of any applicable law, the Design/Builder shall notify the Owner in writing. Neither the Design/Builder nor the Architect shall be obligated to perform any act which either believes will violate any applicable law.

1.2.3 Nothing contained in this Part 2 Agreement shall create a contractual relationship between the Owner and any person or entity other than the Design/Builder.

1.3 OWNERSHIP AND USE OF DOCUMENTS

1.3.1 Drawings, specifications, and other documents and electronic data furnished by the Design/Builder are instruments of service. The Design/Builder's Architect and other providers of professional services shall retain all common law, statutory and other reserved rights, including copyright in those instruments of service furnished by them. Drawings, specifications, and other documents and electronic data are furnished for use solely with respect to this Part 2 Agreement. The Owner shall be permitted to retain copies, including reproducible copies, of the drawings, specifications, and other documents and electronic data furnished by the

Design/Builder for information and reference in connection with the Project except as provided in Subparagraphs 1.3.2 and 1.3.3.

1.3.2 Drawings, specifications, and other documents and electronic data furnished by the Design/Builder shall not be used by the Owner or others on other projects, for additions to this Project or for completion of this Project by others, except by agreement in writing and with appropriate compensation to the Design/Builder, unless the Design/Builder is adjudged to be in default under this Part 2 Agreement or under any other subsequently executed agreement.

1.3.3 If the Design/Builder defaults in the Design/Builder's obligations to the Owner, the Architect shall grant a license to the Owner to use the drawings, specifications, and other documents and electronic data furnished by the Architect to the Design/Builder for the completion of the Project, conditioned upon the Owner's execution of an agreement to cure the Design/Builder's default in payment to the Architect for services previously performed and to indemnify the Architect with regard to claims arising from such reuse without the Architect's professional involvement.

1.3.4 Submission or distribution of the Design/Builder's documents to meet official regulatory requirements or for similar purposes in connection with the Project is not to be construed as publication in derogation of the rights reserved in Subparagraph 1.3.1.

ARTICLE 2
OWNER

2.1 The Owner shall designate a representative authorized to act on the Owner's behalf with respect to the Project. The Owner or such authorized representative shall examine documents submitted by the Design/Builder and shall render decisions in a timely manner and in accordance with the schedule accepted by the Owner. The Owner may obtain independent review of the Contract Documents by a separate architect, engineer, contractor or cost estimator under contract to or employed by the Owner. Such independent review shall be undertaken at the Owner's expense in a timely manner and shall not delay the orderly progress of the Work.

2.2 The Owner may appoint an on-site project representative to observe the Work and to have such other responsibilities as the Owner and the Design/Builder agree in writing.

2.3 The Owner shall cooperate with the Design/Builder in securing building and other permits, licenses and inspections. The Owner shall not be required to pay the fees for such permits, licenses and inspections unless the cost of such fees is excluded from the Design/Builder's Proposal.

2.4 The Owner shall furnish services of land surveyors,

geotechnical engineers and other consultants for subsoil, air and water conditions, in addition to those provided under the Part 1 Agreement, when such services are deemed necessary by the Design/Builder to properly carry out the design services required by this Part 2 Agreement.

2.5 The Owner shall disclose, to the extent known to the Owner, the results and reports of prior tests, inspections or investigations conducted for the Project involving: structural or mechanical systems; chemical, air and water pollution; hazardous materials; or other environmental and subsurface conditions. The Owner shall disclose all information known to the Owner regarding the presence of pollutants at the Project's site.

2.6 The Owner shall furnish all legal, accounting and insurance counseling services as may be necessary at any time for the Project, including such auditing services as the Owner may require to verify the Design/Builder's Applications for Payment.

2.7 Those services, information, surveys and reports required by Paragraphs 2.4 through 2.6 which are within the Owner's control shall be furnished at the Owner's expense, and the Design/Builder shall be entitled to rely upon the accuracy and completeness thereof, except to the extent the Owner advises the Design/Builder to the contrary in writing.

2.8 If the Owner requires the Design/Builder to maintain any special insurance coverage, policy, amendment, or rider, the Owner shall pay the additional cost thereof, except as otherwise stipulated in this Part 2 Agreement.

2.9 If the Owner observes or otherwise becomes aware of a fault or defect in the Work or nonconformity with the Design/Builder's Proposal or the Construction Documents, the Owner shall give prompt written notice thereof to the Design/Builder.

2.10 The Owner shall, at the request of the Design/ Builder, prior to execution of this Part 2 Agreement and promptly upon request thereafter, furnish to the Design/Builder reasonable evidence that financial arrangements have been made to fulfill the Owner's obligations under the Contract.

2.11 The Owner shall communicate with persons or entities employed or retained by the Design/Builder through the Design/Builder, unless otherwise directed by the Design/Builder.

ARTICLE 3

DESIGN/BUILDER

3.1 SERVICES AND RESPONSIBILITIES

3.1.1 Design services required by this Part 2 Agreement shall be performed by qualified architects and other design professionals. The contractual obligations of such professional persons or entities are undertaken and performed in the interest of the Design/Builder.

3.1.2 The agreements between the Design/Builder and

the persons or entities identified in this Part 2 Agreement, and any subsequent modifications, shall be in writing. These agreements, including financial arrangements with respect to this Project, shall be promptly and fully disclosed to the Owner upon request.

3.1.3 The Design/Builder shall be responsible to the Owner for acts and omissions of the Design/Builder's employees, subcontractors and their agents and employees, and other persons, including the Architect and other design professionals, performing any portion of the Design/Builder's obligations under this Part 2 Agreement.

3.2 BASIC SERVICES

3.2.1 The Design/Builder's Basic Services are described below and in Article 14.

3.2.2 The Design/Builder shall designate a representative authorized to act on the Design/Builder's behalf with respect to the Project.

3.2.3 The Design/Builder shall submit Construction Documents for review and approval by the Owner. Construction Documents may include drawings, specifications, and other documents and electronic data setting forth in detail the requirements for construction of the Work, and shall:

.1 be consistent with the intent of the Design/Builder's Proposal;

.2 provide information for the use of those in the building trades; and

.3 include documents customarily required for regulatory agency approvals.

3.2.4 The Design/Builder, with the assistance of the Owner, shall file documents required to obtain necessary approvals of governmental authorities having jurisdiction over the Project.

3.2.5 Unless otherwise provided in the Contract Documents, the Design/Builder shall provide or cause to be provided and shall pay for design services, labor, materials, equipment, tools, construction equipment and machinery, water, heat, utilities, transportation and other facilities and services necessary for proper execution and completion of the Work, whether temporary or permanent and whether or not incorporated or to be incorporated in the Work.

3.2.6 The Design/Builder shall be responsible for all construction means, methods, techniques, sequences and procedures, and for coordinating all portions of the Work under this Part 2 Agreement.

3.2.7 The Design/Builder shall keep the Owner informed of the progress and quality of the Work.

3.2.8 The Design/Builder shall be responsible for correcting Work which does not conform to the Contract Documents.

3.2.9 The Design/Builder warrants to the Owner that materials and equipment furnished under the Contract will be of good quality and new unless otherwise required or permitted by the Contract Documents, that the construction will be free from faults and defects, and that the construction will conform with the requirements of the

Figure 5-5, AIA Document A191

Contract Documents. Construction not conforming to these requirements, including substitutions not properly approved by the Owner, shall be corrected in accordance with Article 9.

3.2.10 The Design/Builder shall pay all sales, consumer, use and similar taxes which had been legally enacted at the time the Design/Builder's Proposal was first submitted to the Owner, and shall secure and pay for building and other permits and governmental fees, licenses and inspections necessary for the proper execution and completion of the Work which are either customarily secured after execution of a contract for construction or are legally required at the time the Design/Builder's Proposal was first submitted to the Owner.

3.2.11 The Design/Builder shall comply with and give notices required by laws, ordinances, rules, regulations and lawful orders of public authorities relating to the Project.

3.2.12 The Design/Builder shall pay royalties and license fees for patented designs, processes or products. The Design/Builder shall defend suits or claims for infringement of patent rights and shall hold the Owner harmless from loss on account thereof, but shall not be responsible for such defense or loss when a particular design, process or product of a particular manufacturer is required by the Owner. However, if the Design/Builder has reason to believe the use of a required design, process or product is an infringement of a patent, the Design/Builder shall be responsible for such loss unless such information is promptly furnished to the Owner.

3.2.13 The Design/Builder shall keep the premises and surrounding area free from accumulation of waste materials or rubbish caused by operations under this Part 2 Agreement. At the completion of the Work, the Design/Builder shall remove from the site waste materials, rubbish, the Design/Builder's tools, construction equipment, machinery, and surplus materials.

3.2.14 The Design/Builder shall notify the Owner when the Design/Builder believes that the Work or an agreed upon portion thereof is substantially completed. If the Owner concurs, the Design/Builder shall issue a Certificate of Substantial Completion which shall establish the Date of Substantial Completion, shall state the responsibility of each party for security, maintenance, heat, utilities, damage to the Work and insurance, shall include a list of items to be completed or corrected and shall fix the time within which the Design/Builder shall complete items listed therein. Disputes between the Owner and Design/Builder regarding the Certificate of Substantial Completion shall be resolved in accordance with provisions of Article 10.

3.2.15 The Design/Builder shall maintain at the site for the Owner one record copy of the drawings, specifications, product data, samples, shop drawings, Change Orders and other modifications, in good order and regularly updated to record the completed construction. These shall be delivered to the Owner upon completion of construction and prior to final payment.

3.3   ADDITIONAL SERVICES

3.3.1 The services described in this Paragraph 3.3 are not included in Basic Services unless so identified in Article 14, and they shall be paid for by the Owner as provided in this Part 2 Agreement, in addition to the compensation for Basic Services. The services described in this Paragraph 3.3 shall be provided only if authorized or confirmed in writing by the Owner.

3.3.2 Making revisions in drawings, specifications, and other documents or electronic data when such revisions are required by the enactment or revision of codes, laws or regulations subsequent to the preparation of such documents or electronic data.

3.3.3 Providing consultation concerning replacement of Work damaged by fire or other cause during construction, and furnishing services required in connection with the replacement of such Work.

3.3.4 Providing services in connection with a public hearing, arbitration proceeding or legal proceeding, except where the Design/Builder is a party thereto.

3.3.5 Providing coordination of construction performed by the Owner's own forces or separate contractors employed by the Owner, and coordination of services required in connection with construction performed and equipment supplied by the Owner.

3.3.6 Preparing a set of reproducible record documents or electronic data showing significant changes in the Work made during construction.

3.3.7 Providing assistance in the utilization of equipment or systems such as preparation of operation and maintenance manuals, training personnel for operation and maintenance, and consultation during operation.

ARTICLE 4

TIME

4.1 Unless otherwise indicated, the Owner and the Design/Builder shall perform their respective obligations as expeditiously as is consistent with reasonable skill and care and the orderly progress of the Project.

4.2 Time limits stated in the Contract Documents are of the essence. The Work to be performed under this Part 2 Agreement shall commence upon receipt of a notice to proceed unless otherwise agreed and, subject to authorized Modifications, Substantial Completion shall be achieved on or before the date established in Article 14.

4.3 Substantial Completion is the stage in the progress of the Work when the Work or designated portion thereof is sufficiently complete in accordance with the Contract Documents so the Owner can occupy or utilize the Work for its intended use.

4.4 Based on the Design/Builder's Proposal, a construction schedule shall be provided consistent with Paragraph 4.2 above.

4.5 If the Design/Builder is delayed at any time in the progress of the Work by an act or neglect of the Owner, Owner's employees, or separate contractors employed by the Owner, or by changes ordered in the Work, or by labor disputes, fire, unusual delay in deliveries, adverse weather conditions not reasonably anticipatable, unavoidable casualties or other causes beyond the Design/Builder's control, or by delay authorized by the Owner pending arbitration, or by other causes which the Owner and Design/Builder agree may justify delay, then the Contract Time shall be reasonably extended by Change Order.

ARTICLE 5
PAYMENTS

5.1 PROGRESS PAYMENTS

5.1.1 The Design/Builder shall deliver to the Owner itemized Applications for Payment in such detail as indicated in Article 14.

5.1.2 Within ten (10) days of the Owner's receipt of a properly submitted and correct Application for Payment, the Owner shall make payment to the Design/Builder.

5.1.3 The Application for Payment shall constitute a representation by the Design/Builder to the Owner that the design and construction have progressed to the point indicated, the quality of the Work covered by the application is in accordance with the Contract Documents, and the Design/Builder is entitled to payment in the amount requested.

5.1.4 Upon receipt of payment from the Owner, the Design/Builder shall promptly pay the Architect, other design professionals and each contractor the amount to which each is entitled in accordance with the terms of their respective contracts.

5.1.5 The Owner shall have no obligation under this Part 2 Agreement to pay or to be responsible in any way for payment to the Architect, another design professional or a contractor performing portions of the Work.

5.1.6 Neither progress payment nor partial or entire use or occupancy of the Project by the Owner shall constitute an acceptance of Work not in accordance with the Contract Documents.

5.1.7 The Design/Builder warrants that title to all construction covered by an Application for Payment will pass to the Owner no later than the time of payment. The Design/Builder further warrants that upon submittal of an Application for Payment all construction for which payments have been received from the Owner shall be free and clear of liens, claims, security interests or encumbrances in favor of the Design/Builder or any other person or entity performing construction at the site or furnishing materials or equipment relating to the construction.

5.1.8 At the time of Substantial Completion, the Owner shall pay the Design/Builder the retainage, if any, less the reasonable cost to correct or complete incorrect or incomplete Work. Final payment of such withheld sum shall be made upon correction or completion of such Work.

5.2 FINAL PAYMENT

5.2.1 Neither final payment nor amounts retained, if any, shall become due until the Design/Builder submits to the Owner: (1) an affidavit that payrolls, bills for materials and equipment, and other indebtedness connected with the Work for which the Owner or Owner's property might be responsible or encumbered (less amounts withheld by the Owner) have been paid or otherwise satisfied; (2) a certificate evidencing that insurance required by the Contract Documents to remain in force after final payment is currently in effect and will not be canceled or allowed to expire until at least 30 days' prior written notice has been given to the Owner; (3) a written statement that the Design/Builder knows of no substantial reason that the insurance will not be renewable to cover the period required by the Contract Documents; (4) consent of surety, if any, to final payment; and (5) if required by the Owner, other data establishing payment or satisfaction of obligations, such as receipts, releases and waivers of liens, claims, security interests or encumbrances arising out of the Contract, to the extent and in such form as may be designated by the Owner. If a contractor or other person or entity entitled to assert a lien against the Owner's property refuses to furnish a release or waiver required by the Owner, the Design/Builder may furnish a bond satisfactory to the Owner to indemnify the Owner against such lien. If such lien remains unsatisfied after payments are made, the Design/Builder shall indemnify the Owner for all loss and cost, including reasonable attorneys' fees incurred as a result of such lien.

5.2.2 When the Work has been completed and the contract fully performed, the Design/Builder shall submit a final application for payment to the Owner, who shall make final payment within 30 days of receipt.

5.2.3 The making of final payment shall constitute a waiver of claims by the Owner except those arising from:

  .1 liens, claims, security interests or encumbrances arising out of the Contract and unsettled;

  .2 failure of the Work to comply with the requirements of the Contract Documents; or

  .3 terms of special warranties required by the Contract Documents.

5.2.4 Acceptance of final payment shall constitute a waiver of all claims by the Design/Builder except those previously made in writing and identified by the Design/Builder as unsettled at the time of final Application for Payment.

5.3 INTEREST PAYMENTS

5.3.1 Payments due the Design/Builder under this Part 2 Agreement which are not paid when due shall bear interest from the date due at the rate specified in Article 13, or in the absence of a specified rate, at the legal rate prevailing where the Project is located.

ARTICLE 6
PROTECTION OF PERSONS AND PROPERTY

6.1 The Design/Builder shall be responsible for initiating, maintaining and providing supervision of all safety precautions and programs in connection with the performance of this Part 2 Agreement.

Figure 5-5, AIA Document A191

6.2 The Design/Builder shall take reasonable precautions for the safety of, and shall provide reasonable protection to prevent damage, injury or loss to: (1) employees on the Work and other persons who may be affected thereby; (2) the Work and materials and equipment to be incorporated therein, whether in storage on or off the site, under care, custody, or control of the Design/Builder or the Design/Builder's contractors; and (3) other property at or adjacent thereto, such as trees, shrubs, lawns, walks, pavements, roadways, structures and utilities not designated for removal, relocation or replacement in the course of construction.

6.3 The Design/Builder shall give notices and comply with applicable laws, ordinances, rules, regulations and lawful orders of public authorities bearing on the safety of persons or property or their protection from damage, injury or loss.

6.4 The Design/Builder shall promptly remedy damage and loss (other than damage or loss insured under property insurance provided or required by the Contract Documents) to property at the site caused in whole or in part by the Design/Builder, a contractor of the Design/Builder or anyone directly or indirectly employed by any of them, or by anyone for whose acts they may be liable.

ARTICLE 7
INSURANCE AND BONDS

7.1 DESIGN/BUILDER'S LIABILITY INSURANCE

7.1.1 The Design/Builder shall purchase from and maintain, in a company or companies lawfully authorized to do business in the jurisdiction in which the Project is located, such insurance as will protect the Design/Builder from claims set forth below which may arise out of or result from operations under this Part 2 Agreement by the Design/Builder or by a contractor of the Design/Builder, or by anyone directly or indirectly employed by any of them, or by anyone for whose acts any of them may be liable:

  .1 claims under workers' compensation, disability benefit and other similar employee benefit laws that are applicable to the Work to be performed;

  .2 claims for damages because of bodily injury, occupational sickness or disease, or death of the Design/Builder's employees;

  .3 claims for damages because of bodily injury, sickness or disease, or death of persons other than the Design/Builder's employees;

  .4 claims for damages covered by usual personal injury liability coverage which are sustained (1) by a person as a result of an offense directly or indirectly related to employment of such person by the Design/Builder or (2) by another person;

  .5 claims for damages, other than to the Work itself, because of injury to or destruction of tangible property, including loss of use resulting therefrom;

  .6 claims for damages because of bodily injury, death of a person or property damage arising out

of ownership, maintenance or use of a motor vehicle; and

  .7 claims involving contractual liability insurance applicable to the Design/Builder's obligations under Paragraph 11.5.

7.1.2 The insurance required by Subparagraph 7.1.1 shall be written for not less than limits of liability specified in this Part 2 Agreement or required by law, whichever coverage is greater. Coverages, whether written on an occurrence or claims-made basis, shall be maintained without interruption from date of commencement of the Work until date of final payment and termination of any coverage required to be maintained after final payment.

7.1.3 Certificates of insurance acceptable to the Owner shall be delivered to the Owner immediately after execution of this Part 2 Agreement. These certificates and the insurance policies required by this Paragraph 7.1 shall contain a provision that coverages afforded under the policies will not be canceled or allowed to expire until at least 30 days' prior written notice has been given to the Owner. If any of the foregoing insurance coverages are required to remain in force after final payment, an additional certificate evidencing continuation of such coverage shall be submitted with the application for final payment. Information concerning reduction of coverage shall be furnished by the Design/Builder with reasonable promptness in accordance with the Design/Builder's information and belief.

7.2 OWNER'S LIABILITY INSURANCE

7.2.1 The Owner shall be responsible for purchasing and maintaining the Owner's usual liability insurance. Optionally, the Owner may purchase and maintain other insurance for self-protection against claims which may arise from operations under this Part 2 Agreement. The Design/Builder shall not be responsible for purchasing and maintaining this optional Owner's liability insurance unless specifically required by the Contract Documents.

7.3 PROPERTY INSURANCE

7.3.1 Unless otherwise provided under this Part 2 Agreement, the Owner shall purchase and maintain, in a company or companies authorized to do business in the jurisdiction in which the principal improvements are to be located, property insurance upon the Work to the full insurable value thereof on a replacement cost basis without optional deductibles. Such property insurance shall be maintained, unless otherwise provided in the Contract Documents or otherwise agreed in writing by all persons and entities who are beneficiaries of such insurance, until final payment has been made or until no person or entity other than the Owner has an insurable interest in the property required by this Paragraph 7.3 to be insured, whichever is earlier. This insurance shall include interests of the Owner, the Design/Builder, and their respective contractors and subcontractors in the Work.

7.3.2 Property insurance shall be on an all-risk policy form and shall insure against the perils of fire and extended coverage and physical loss or damage including, without duplication of coverage, theft, vandalism, malicious mischief, collapse, falsework, temporary buildings and debris removal including demolition occasioned by enforcement of any applicable legal requirements, and shall cover reasonable compensation for the services and expenses of the Design/Builder's Architect and other professionals required as a result of such insured loss. Coverage for other perils shall not be required unless otherwise provided in the Contract Documents.

7.3.3 If the Owner does not intend to purchase such property insurance required by this Part 2 Agreement and with all of the coverages in the amount described above, the Owner shall so inform the Design/Builder prior to commencement of the construction. The Design/Builder may then effect insurance which will protect the interests of the Design/Builder and the Design/Builder's contractors in the construction, and by appropriate Change Order the cost thereof shall be charged to the Owner. If the Design/Builder is damaged by the failure or neglect of the Owner to purchase or maintain insurance as described above, then the Owner shall bear all reasonable costs properly attributable thereto.

7.3.4 Unless otherwise provided, the Owner shall purchase and maintain such boiler and machinery insurance required by this Part 2 Agreement or by law, which shall specifically cover such insured objects during installation and until final acceptance by the Owner. This insurance shall include interests of the Owner, the Design/Builder, the Design/Builder's contractors and subcontractors in the Work, and the Design/Builder's Architect and other design professionals. The Owner and the Design/Builder shall be named insureds.

7.3.5 A loss insured under the Owner's property insurance shall be adjusted by the Owner as fiduciary and made payable to the Owner as fiduciary for the insureds, as their interests may appear, subject to requirements of any applicable mortgagee clause and of Subparagraph 7.3.10. The Design/Builder shall pay contractors their shares of insurance proceeds received by the Design/Builder, and by appropriate agreement, written where legally required for validity, shall require contractors to make payments to their subcontractors in similar manner.

7.3.6 Before an exposure to loss may occur, the Owner shall file with the Design/Builder a copy of each policy that includes insurance coverages required by this Paragraph 7.3. Each policy shall contain all generally applicable conditions, definitions, exclusions and endorsements related to this Project. Each policy shall contain a provision that the policy will not be canceled or allowed to expire until at least 30 days' prior written notice has been given to the Design/Builder.

7.3.7 If the Design/Builder requests in writing that insurance for risks other than those described herein or for other special hazards be included in the property insurance policy, the Owner shall, if possible, obtain such insurance, and the cost thereof shall be charged to the Design/Builder by appropriate Change Order.

7.3.8 The Owner and the Design/Builder waive all rights against each other and the Architect and other design professionals, contractors, subcontractors, agents and employees, each of the other, for damages caused by fire or other perils to the extent covered by property insurance obtained pursuant to this Paragraph 7.3 or other property insurance applicable to the Work, except such rights as they may have to proceeds of such insurance held by the Owner as trustee. The Owner or Design/Builder, as appropriate, shall require from contractors and subcontractors by appropriate agreements, written where legally required for validity, similar waivers each in favor of other parties enumerated in this Paragraph 7.3. The policies shall provide such waivers of subrogation by endorsement or otherwise. A waiver of subrogation shall be effective as to a person or entity even though that person or entity would otherwise have a duty of indemnification, contractual or otherwise, did not pay the insurance premium directly or indirectly, and whether or not the person or entity had an insurable interest in the property damaged.

7.3.9 If required in writing by a party in interest, the Owner as trustee shall, upon occurrence of an insured loss, give bond for proper performance of the Owner's duties. The cost of required bonds shall be charged against proceeds received as fiduciary. The Owner shall deposit in a separate account proceeds so received, which the Owner shall distribute in accordance with such agreement as the parties in interest may reach, or in accordance with an arbitration award in which case the procedure shall be as provided in Article 10. If after such loss no other special agreement is made, replacement of damaged Work shall be covered by appropriate Change Order.

7.3.10 The Owner as trustee shall have power to adjust and settle a loss with insurers unless one of the parties in interest shall object in writing, within five (5) days after occurrence of loss to the Owner's exercise of this power; if such objection be made, the parties shall enter into dispute resolution under procedures provided in Article 10. If distribution of insurance proceeds by arbitration is required, the arbitrators will direct such distribution.

7.3.11 Partial occupancy or use prior to Substantial Completion shall not commence until the insurance company or companies providing property insurance have consented to such partial occupancy or use by endorsement or otherwise. The Owner and the Design/Builder shall take reasonable steps to obtain consent of the insurance company or companies and shall not, without mutual written consent, take any action with respect to partial occupancy or use that would cause cancellation, lapse or reduction of coverage.

7.4 LOSS OF USE INSURANCE

7.4.1 The Owner, at the Owner's option, may purchase and maintain such insurance as will insure the Owner against loss of use of the Owner's property due to fire or other hazards, however caused. The Owner waives all rights of action against the Design/Builder for loss of use of the Owner's property, including consequential losses due to fire or other hazards, however caused.

222

Figure 5-5, AIA Document A191

ARTICLE 8
CHANGES IN THE WORK

8.1 CHANGES

8.1.1 Changes in the Work may be accomplished after execution of this Part 2 Agreement, without invalidating this Part 2 Agreement, by Change Order, Construction Change Directive, or order for a minor change in the Work, subject to the limitations stated in the Contract Documents.

8.1.2 A Change Order shall be based upon agreement between the Owner and the Design/Builder; a Construction Change Directive may be issued by the Owner without the agreement of the Design/Builder; an order for a minor change in the Work may be issued by the Design/Builder alone.

8.1.3 Changes in the Work shall be performed under applicable provisions of the Contract Documents, and the Design/Builder shall proceed promptly, unless otherwise provided in the Change Order, Construction Change Directive, or order for a minor change in the Work.

8.1.4 If unit prices are stated in the Contract Documents or subsequently agreed upon, and if quantities originally contemplated are so changed in a proposed Change Order or Construction Change Directive that application of such unit prices to quantities of Work proposed will cause substantial inequity to the Owner or the Design/Builder, the applicable unit prices shall be equitably adjusted.

8.2 CHANGE ORDERS

8.2.1 A Change Order is a written instrument prepared by the Design/Builder and signed by the Owner and the Design/Builder, stating their agreement upon all of the following:

   .1 a change in the Work;

   .2 the amount of the adjustment, if any, in the Contract Sum; and

   .3 the extent of the adjustment, if any, in the Contract Time.

8.2.2 If the Owner requests a proposal for a change in the Work from the Design/Builder and subsequently elects not to proceed with the change, a Change Order shall be issued to reimburse the Design/Builder for any costs incurred for estimating services, design services or preparation of proposed revisions to the Contract Documents.

8.3 CONSTRUCTION CHANGE DIRECTIVES

8.3.1 A Construction Change Directive is a written order prepared and signed by the Owner, directing a change in the Work prior to agreement on adjustment, if any, in the Contract Sum or Contract Time, or both.

8.3.2 Except as otherwise agreed by the Owner and the Design/Builder, the adjustment to the Contract Sum shall be determined on the basis of reasonable expenditures and savings of those performing the Work attributable to the change, including the expenditures for design services and revisions to the Contract Documents. In case of an increase in the Contract Sum, the cost shall include a reasonable allowance for overhead and profit. In such case, the Design/Builder shall keep and present an itemized accounting together with appropriate supporting data for inclusion in a Change Order. Unless otherwise provided in the Contract Documents, costs for these purposes shall be limited to the following:

   .1 costs of labor, including social security, old age and unemployment insurance, fringe benefits required by agreement or custom, and workers' compensation insurance;

   .2 costs of materials, supplies and equipment, including cost of transportation, whether incorporated or consumed;

   .3 rental costs of machinery and equipment exclusive of hand tools, whether rented from the Design/Builder or others;

   .4 costs of premiums for all bonds and insurance permit fees, and sales, use or similar taxes;

   .5 additional costs of supervision and field office personnel directly attributable to the change; and fees paid to the Architect, engineers and other professionals.

8.3.3 Pending final determination of cost to the Owner, amounts not in dispute may be included in Applications for Payment. The amount of credit to be allowed by the Design/Builder to the Owner for deletion or change which results in a net decrease in the Contract Sum will be actual net cost. When both additions and credits covering related Work or substitutions are involved in a change, the allowance for overhead and profit shall be figured on the basis of the net increase, if any, with respect to that change.

8.3.4 When the Owner and the Design/Builder agree upon the adjustments in the Contract Sum and Contract Time, such agreement shall be effective immediately and shall be recorded by preparation and execution of an appropriate Change Order.

8.4 MINOR CHANGES IN THE WORK

8.4.1 The Design/Builder shall have authority to make minor changes in the Construction Documents and construction consistent with the intent of the Contract Documents when such minor changes do not involve adjustment in the Contract Sum or extension of the Contract Time. The Design/Builder shall promptly inform the Owner, in writing, of minor changes in the Construction Documents and construction.

8.5 CONCEALED CONDITIONS

8.5.1 If conditions are encountered at the site which are (1) subsurface or otherwise concealed physical conditions which differ materially from those indicated in the Contract Documents, or (2) unknown physical conditions of an unusual nature which differ materially from those ordinarily found to exist and generally recognized as inherent in construction activities of the character provided for in the Contract Documents, then notice by the observing party shall be given to the other party promptly before conditions are disturbed and in no event later than 21 days after first observance of the conditions. The Contract Sum shall be equitably adjusted for such concealed or unknown conditions by Change Order upon claim by either party made within 21 days after the claimant becomes aware of the conditions.

8.6 REGULATORY CHANGES

8.6.1 The Design/Builder shall be compensated for changes in the construction necessitated by the enactment or revision of codes, laws or regulations subsequent to the submission of the Design/Builder's Proposal.

ARTICLE 9

CORRECTION OF WORK

9.1 The Design/Builder shall promptly correct Work rejected by the Owner or known by the Design/Builder to be defective or failing to conform to the requirements of the Contract Documents, whether observed before or after Substantial Completion and whether or not fabricated, installed or completed. The Design/Builder shall bear costs of correcting such rejected Work, including additional testing and inspections.

9.2 If, within one (1) year after the date of Substantial Completion of the Work or, after the date for commencement of warranties established in a written agreement between the Owner and the Design/Builder, or by terms of an applicable special warranty required by the Contract Documents, any of the Work is found to be not in accordance with the requirements of the Contract Documents, the Design/Builder shall correct it promptly after receipt of a written notice from the Owner to do so unless the Owner has previously given the Design/Builder a written acceptance of such condition.

9.3 Nothing contained in this Article 9 shall be construed to establish a period of limitation with respect to other obligations which the Design/Builder might have under the Contract Documents. Establishment of the time period of one (1) year as described in Subparagraph 9.2 relates only to the specific obligation of the Design/Builder to correct the Work, and has no relationship to the time within which the obligation to comply with the Contract Documents may be sought to be enforced, nor to the time within which proceedings may be commenced to establish the Design/Builder's liability with respect to the Design/Builder's obligations other than specifically to correct the Work.

9.4 If the Design/Builder fails to correct nonconforming Work as required or fails to carry out Work in accordance with the Contract Documents, the Owner, by written order signed personally or by an agent specifically so empowered by the Owner in writing, may order the Design/Builder to stop the Work, or any portion thereof, until the cause for such order has been eliminated; however, the Owner's right to stop the Work shall not give rise to a duty on the part of the Owner to exercise the right for benefit of the Design/Builder or other persons or entities.

9.5 If the Design/Builder defaults or neglects to carry out the Work in accordance with the Contract Documents and fails within seven (7) days after receipt of written notice from the Owner to commence and continue correction of such default or neglect with diligence and promptness, the Owner may give a second written notice to the Design/Builder and, seven (7) days following receipt by the Design/Builder of that second written notice and without prejudice to other remedies the Owner may have, correct such deficiencies. In such case an appropriate Change Order shall be issued deducting from payments then or thereafter due the Design/Builder, the costs of correcting such deficiencies. If the payments then or thereafter due the Design/Builder are not sufficient to cover the amount of the deduction, the Design/Builder shall pay the difference to the Owner. Such action by the Owner shall be subject to dispute resolution procedures as provided in Article 10.

ARTICLE 10

DISPUTE RESOLUTION— MEDIATION AND ARBITRATION

10.1 Claims, disputes or other matters in question between the parties to this Part 2 Agreement arising out of or relating to this Part 2 Agreement or breach thereof shall be subject to and decided by mediation or arbitration. Such mediation or arbitration shall be conducted in accordance with the Construction Industry Mediation or Arbitration Rules of the American Arbitration Association currently in effect.

10.2 In addition to and prior to arbitration, the parties shall endeavor to settle disputes by mediation. Demand for mediation shall be filed in writing with the other party to this Part 2 Agreement and with the American Arbitration Association. A demand for mediation shall be made within a reasonable time after the claim, dispute or other matter in question has arisen. In no event shall the demand for mediation be made after the date when institution of legal or equitable proceedings based on such claim, dispute or other matter in question would be barred by the applicable statutes of repose or limitations.

10.3 Demand for arbitration shall be filed in writing with the other party to this Part 2 Agreement and with the American Arbitration Association. A demand for arbitration shall be made within a reasonable time after the claim, dispute or other matter in question has arisen. In no event shall the demand for arbitration be made after the date when institution of legal or equitable proceedings based on such claim, dispute or other matter in question would be barred by the applicable statutes of repose or limitations.

10.4 An arbitration pursuant to this Article may be joined with an arbitration involving common issues of law or fact between the Design/Builder and any person or entity with whom the Design/Builder has a contractual obligation to arbitrate disputes. No other arbitration arising out of or relating to this Part 2 Agreement shall include, by consolidation, joinder or in any other manner, an additional person or entity not a party to this Part 2 Agreement or not a party to an agreement with the Design/Builder, except by written consent containing a

Figure 5-5, AIA Document A191

specific reference to this Part 2 Agreement signed by the Owner, the Design/Builder and any other person or entities sought to be joined. Consent to arbitration involving an additional person or entity shall not constitute consent to arbitration of any claim, dispute or other matter in question not described in the written consent or with a person or entity not named or described therein. The foregoing agreement to arbitrate and other agreements to arbitrate with an additional person or entity duly consented to by the parties to this Part 2 Agreement shall be specifically enforceable in accordance with applicable law in any court having jurisdiction thereof.

10.5 The award rendered by the arbitrator or arbitrators shall be final, and judgment may be entered upon it in accordance with applicable law in any court having jurisdiction thereof.

ARTICLE 11

MISCELLANEOUS PROVISIONS

11.1 Unless otherwise provided, this Part 2 Agreement shall be governed by the law of the place where the Project is located.

11.2 SUBCONTRACTS

11.2.1 The Design/Builder, as soon as practicable after execution of this Part 2 Agreement, shall furnish to the Owner in writing the names of the persons or entities the Design/Builder will engage as contractors for the Project.

11.3 WORK BY OWNER OR OWNER'S CONTRACTORS

11.3.1 The Owner reserves the right to perform construction or operations related to the Project with the Owner's own forces, and to award separate contracts in connection with other portions of the Project or other construction or operations on the site under conditions of insurance and waiver of subrogation identical to the provisions of this Part 2 Agreement. If the Design/Builder claims that delay or additional cost is involved because of such action by the Owner, the Design/Builder shall assert such claims as provided in Subparagraph 11.4.

11.3.2 The Design/Builder shall afford the Owner's separate contractors reasonable opportunity for introduction and storage of their materials and equipment and performance of their activities and shall connect and coordinate the Design/Builder's construction and operations with theirs as required by the Contract Documents.

11.3.3 Costs caused by delays or by improperly timed activities or defective construction shall be borne by the party responsible therefor.

11.4 CLAIMS FOR DAMAGES

11.4.1 If either party to this Part 2 Agreement suffers injury or damage to person or property because of an act or omission of the other party, of any of the other party's employees or agents, or of others for whose acts such party is legally liable, written notice of such injury or damage, whether or not insured, shall be given to the

other party within a reasonable time not exceeding 21 days after first observance. The notice shall provide sufficient detail to enable the other party to investigate the matter. If a claim of additional cost or time related to this claim is to be asserted, it shall be filed in writing.

11.5 INDEMNIFICATION

11.5.1 To the fullest extent permitted by law, the Design/Builder shall indemnify and hold harmless the Owner, Owner's consultants, and agents and employees of any of them from and against claims, damages, losses and expenses, including but not limited to attorneys' fees, arising out of or resulting from performance of the Work, provided that such claim, damage, loss or expense is attributable to bodily injury, sickness, disease or death, or to injury to or destruction of tangible property (other than the Work itself) including loss of use resulting therefrom, but only to the extent caused in whole or in part by negligent acts or omissions of the Design/Builder, anyone directly or indirectly employed by the Design/Builder or anyone for whose acts the Design/Builder may be liable, regardless of whether or not such claim, damage, loss or expense is caused in part by a party indemnified hereunder. Such obligation shall not be construed to negate, abridge, or reduce other rights or obligations of indemnity which would otherwise exist as to a party or person described in this Paragraph 11.5.

11.5.2 In claims against any person or entity indemnified under this Paragraph 11.5 by an employee of the Design/Builder, anyone directly or indirectly employed by the Design/Builder or anyone for whose acts the Design/Builder may be liable, the indemnification obligation under this Paragraph 11.5 shall not be limited by a limitation on amount or type of damages, compensation or benefits payable by or for the Design/Builder under workers' compensation acts, disability benefit acts or other employee benefit acts.

11.6 SUCCESSORS AND ASSIGNS

11.6.1 The Owner and Design/Builder, respectively, bind themselves, their partners, successors, assigns and legal representatives to the other party to this Part 2 Agreement and to the partners, successors and assigns of such other party with respect to all covenants of this Part 2 Agreement. Neither the Owner nor the Design/Builder shall assign this Part 2 Agreement without the written consent of the other. The Owner may assign this Part 2 Agreement to any institutional lender providing construction financing, and the Design/Builder agrees to execute all consents reasonably required to facilitate such an assignment. If either party makes such an assignment, that party shall nevertheless remain legally responsible for all obligations under this Part 2 Agreement, unless otherwise agreed by the other party.

11.7 TERMINATION OF PROFESSIONAL DESIGN SERVICES

11.7.1 Prior to termination of the services of the Architect or any other design professional designated in this Part 2 Agreement, the Design/Builder shall identify to the Owner in writing another architect or other design professional with respect to whom the Owner has no reasonable objection, who will provide the services originally to have been provided by the Architect or other design professional whose services are being terminated.

11.8 EXTENT OF AGREEMENT

11.8.1 This Part 2 Agreement represents the entire agreement between the Owner and the Design/Builder and supersedes prior negotiations, representations or agreements, either written or oral. This Part 2 Agreement may be amended only by written instrument and signed by both the Owner and the Design/Builder.

ARTICLE 12

TERMINATION OF THE AGREEMENT

12.1 TERMINATION BY THE OWNER

12.1.1 This Part 2 Agreement may be terminated by the Owner upon 14 days' written notice to the Design/Builder in the event that the Project is abandoned. If such termination occurs, the Owner shall pay the Design/Builder for Work completed and for proven loss sustained upon materials, equipment, tools, and construction equipment and machinery, including reasonable profit and applicable damages.

12.1.2 If the Design/Builder defaults or persistently fails or neglects to carry out the Work in accordance with the Contract Documents or fails to perform the provisions of this Part 2 Agreement, the Owner may give written notice that the Owner intends to terminate this Part 2 Agreement. If the Design/Builder fails to correct the defaults, failure or neglect within seven (7) days after being given notice, the Owner may then give a second written notice and, after an additional seven (7) days, the Owner may without prejudice to any other remedy terminate the employment of the Design/Builder and take possession of the site and of all materials, equipment, tools and construction equipment and machinery thereon owned by the Design/Builder and finish the Work by whatever method the Owner may deem expedient. If the unpaid balance of the Contract Sum exceeds the expense of finishing the Work and all damages incurred by the Owner, such excess shall be paid to the Design/Builder. If the expense of completing the Work and all damages incurred by the Owner exceeds the unpaid balance, the Design/Builder shall pay the difference to the Owner. This obligation for payment shall survive termination of this Part 2 Agreement.

12.2 TERMINATION BY THE DESIGN/BUILDER

12.2.1 If the Owner fails to make payment when due, the Design/Builder may give written notice of the Design/Builder's intention to terminate this Part 2 Agreement. If the Design/Builder fails to receive payment within seven (7) days after receipt of such notice by the Owner, the Design/Builder may give a second written notice and, seven (7) days after receipt of such second written notice by the Owner, may terminate this Part 2 Agreement and recover from the Owner payment for Work executed and for proven losses sustained upon materials, equipment, tools, and construction equipment and machinery, including reasonable profit and applicable damages.

ARTICLE 13

BASIS OF COMPENSATION

The Owner shall compensate the Design/Builder in accordance with Article 5, Payments, and the other provisions of this Part 2 Agreement as described below.

13.1 COMPENSATION

13.1.1 For the Design/Builder's performance of the Work, as described in Paragraph 3.2 and including any other services listed in Article 14 as part of Basic Services, the Owner shall pay the Design/Builder in current funds the Contract Sum as follows:

Figure 5-5, AIA Document A191

13.1.2 For Additional Services, as described in Paragraph 3.3 and including any other services listed in Article 14 as Additional Services, compensation shall be as follows:

13.2   REIMBURSABLE EXPENSES

13.2.1 Reimbursable Expenses are in addition to the compensation for Basic and Additional Services, and include actual expenditures made by the Design/Builder and the Design/Builder's employees and contractors in the interest of the Project, as follows:

13.2.2 FOR REIMBURSABLE EXPENSES, compensation shall be a multiple of (   ) times the amounts expended.

13.3   INTEREST PAYMENTS

13.3.1 The rate of interest for past due payments shall be as follows:

(Usury laws and requirements under the Federal Truth in Lending Act, similar state and local consumer credit laws and other regulations at the Owner's and Design/ Builder's principal places of business, at the location of the Project and elsewhere may affect the validity of this provision. Specific legal advice should be obtained with respect to deletion, modification or other requirements, such as written disclosures or waivers.)

ARTICLE 14
OTHER CONDITIONS AND SERVICES

14.1   The Basic Services to be performed shall be commenced on
and, subject to authorized adjustments and to delays not caused by the Design/Builder, Substantial Completion shall be
achieved in the Contract Time of   (   ) calendar days.

14.2   The Basic Services beyond those described in Article 3 are as follows:

14.3   Additional Services beyond those described in Article 3 are as follows:

14.4   The Design/Builder shall submit an Application for Payment on the
(   ) day of each month.

A191—1996
Part 2—Page 14

Figure 5-5, AIA Document A191

14.5 The Design/Builder's Proposal includes the following documents:
(List the documents by specific title and date; include any required performance and payment bonds.)

Title   Date

Sample

This Agreement entered into as of the day and year first written above.

OWNER   DESIGN/BUILDER

_____   _____
(Signature)   *(Signature)*

_____   _____
(Printed name and title)   *(Printed name and title)*

AIA CAUTION: You should sign an original AIA document which has this caution printed in red. An original assures that changes will not be obscured as may occur when documents are reproduced.

Index